INTRACEREBRAL HEMORRHAGE

Edited by **Vikas Chaudhary**

Intracerebral Hemorrhage
http://dx.doi.org/10.5772/56977
Edited by Vikas Chaudhary

Contributors

Cathie Sudlow, Kristiina Rannikmae, Gaiqing Wang, Da-Zhi Liu, Stella Tsirka, Shahina Bano, Vikas Chaudhary, Sankalp Gokhale, Michael L. \"Luke\" James

Published by InTech

Janeza Trdine 9, 51000 Rijeka, Croatia

Publishing Process Manager Iva Lipovic

Technical Editor InTech DTP team

Cover InTech Design team

First published September, 2014

Printed in Croatia

Additional hard copies can be obtained from orders@intechopen.com

Intracerebral Hemorrhage, Edited by Vikas Chaudhary

p. cm.

ISBN 978-953-51-1722-3

For EU product safety concerns:
IN TECH d.o.o., Prolaz Marije Krucifikse Kozulić 3, 51000 Rijeka, Croatia,
info@intechopen.com or visit our website at intechopen.com.

Meet the editor

Dr Vikas Chaudhary is an Indian Radiologist who has published in many national and international publications. His other achievements include authoring chapters in other books, and reviewing articles for reputed national and international indexed journals. He was awarded Shri. M.L. Garg Memorial Medal for Best Senior Resident in Radiology at Lady Hardinge Medical College, New Delhi in 2009-2010, and was also awarded second prize for best review article with his contribution "Imaging of the Pancreas: Recent advances", during the Indian Journal of Endocrinology & Metabolism Annual Awards 2011, organised by the Endocrine Society of India at Pune, India. He has been nominated as a specialist member (subject expert) in the recruitment board for appointment of specialists, senior residents and radiographers in various ESIC Model Hospitals, and as an inspecting officer for various radiological equipments. He has received his MBBS and MD (Radiodiagnosis) from reputed Government Medical Colleges of India. During his undergraduation (MBBS), he received a Certificate of Merit in Microbiology.

Contents

Preface

Intracerebral hemorrhage is an important clinical entity encountered in practice. Common causes of intracerebral hemorrhage include hypertension, amyloid angiopathy, trauma, coagulopathy, arteriovenous malformation and underlying tumor. Advances in imaging techniques have helped in better understanding of pathogenesis and the mechanisms of recovery of intracerebral hemorrhage, thereby resulting in marked improvement in its management. I hope that this book on intracerebral hemorrhage will be a useful learning tool for students and clinicians in the field of neuroscience.

We couldn't have produced this book within a year without the stellar author team. I am deeply indebted to all the authors. Their contributions and hard work made this a worthwhile and proud endeavour.

Special thanks goes to Ms. Iva Lipović, publishing process manager for this project, for her tireless and unwavering support of the project.

We are greatly indebted to the production staff of InTech Publisher for their patience, valuable guidance and extreme professionalism.

I would also like to thank my family, friends and colleagues for the understanding and encouragement throughout the project.

Dr. Vikas Chaudhary [MBBS, MD (Radiodiagnosis)],
Specialist and Head, Department of Radiodiagnosis,
Employees' State Insurance Corporation (ESIC) Model Hospital,
Sector-9A,
Gurgaon-122001, Haryana, India

Chapter 1

Intracranial Hemorrhage in the Newborn

Shahina Bano, Vikas Chaudhary,
Umesh Chandra Garga, Sachchidanand Yadav and
Sachin Kumar Singh

Additional information is available at the end of the chapter

http://dx.doi.org/10.5772/58476

1. Introduction

Intracranial hemorrhage (ICH) is a major source of neonatal morbidity and mortality. In full-term infants, it most often occurs during labor as the result of mechanical factors; however, in the pre-term infants it can occur even prior to labor or as late as the second week of life usually as a result of hemodynamic instability. Besides etiology, the location of hemorrhage, clinical presentation and neurological outcome also differs in the term and preterm infants. It is important for the radiologists to provide an accurate anatomic description of the compartment(s) confining the hemorrhage, as correct location may be an indicator to the underlying cause and provide a roadmap to the neurosurgeons if intervention is required. The knowledge of the anatomic compartments is vital for interpreting the imaging findings in case of ICH and formulating a differential diagnosis. Cranial ultrasound is often used as the first imaging modality for newborns. CT is the preferred diagnostic study for evaluation of acute intracranial hemorrhage. MRI is indicated when subarachnoid bleed or posterior fossa hemorrhage is suspected. Prevention of ICH is a subject of great interest in premature newborns. Prenatal prophylaxis and improved obstetric and neonatal care in general markedly reduces the stress to premature fetus and neonate.

2. Etiopathogenesis

Intracranial hemorrhage (ICH) is perhaps the most dramatic manifestation inherited in the birth process. The etiology of ICH differs according to the gestational age of the infant and the site of hemorrhage.

Of all types of ICHs, germinal matrix-intraventricular hemorrhage (GM-IVH) by far is the most common and distinctive pathology in premature infants. The pathogenesis of GM-IVH is multifactorial. It involves a combination of vascular-anatomic immaturity and complex hemodynamic factors. The role of inflammatory and genetic factors is currently under investigation. Germinal matrix (GM) is a highly vascular structure and the source of developing brain cells (neuronal and glial cells). The tissue surrounds the fetal ventricular system and gradually involutes to reside over the body of the caudate between 24 and 28 weeks of gestation, at the level of head of caudate nucleus in the caudothalamic groove between 28 and 34 weeks of gestation, and finally completely regresses and converts into normal cerebral parenchyma by the 36th week of gestation. The capillary network of germinal matrix is composed of high-caliber, thin-walled (deficient in muscularis layer) and immature fragile vessels predisposed to rupture. Furthermore, GM lies within an arterial end zone, and is directly connected to the deep galenic venous system, thereby exposing it to insults of arterial ischemic-reperfusion and to venous congestion.[1,2,3] The rupture hemorrhage of the vulnerable GM requires the coexistence of several intrinsic and extrinsic hemodynamic factors. Premature infants are believed to have impaired cerebral pressure autoregulation (a major intrinsic factor) that renders these infants susceptible to both cerebral hypoperfusion and ischemia at the border zone GM vessels and hence the rupture of fragile germinal matrix vessels. In term infants, a well developed cerebral pressure autoregulation mechanism maintains a relatively constant cerebral blood flow (CBF) across a range of cerebral perfusion pressures.[4] Various extrinsic hemodynamic factors that potentially interfere with the integrity of the vulnerable GM include low CBF (hypotensive events and frank perinatal asphyxia), high CBF (hypertension, bolus fluid infusion, hypercarbia and low hematocrit), fluctuating CBF, and factors causing increased cerebral venous pressure (respiratory distress syndrome, positive pressure ventilation, pneumothorax, or pulmonary hemorrhage).[1,2] Immature deep galenic system, in a preterm infant, which is prone to venous congestion and stasis is another major factor responsible for development of GM-IVH and its complications. Immature cerebral venous system has several vulnerabilities because (i) development of cerebral venous system occurs late in relation to that of the arteries, (ii) there is sequential remodeling and considerable individual variation in the pattern and size of different veins entering the internal cerebral veins, (iii) immature veins have high caliber and thin wall, they branch parallel to the ventricle, hence tend to collapse, (iv) because of relative paucity of superficial cortical veins between 24 and 28 weeks of gestation, most of the cerebral venous drainage is dependent on the deep galenic system that drains GM and most of the white matter, and (v) the periventricular veins, particularly the terminal (thalamostriate) vein, which is the main vein draining the white matter passes directly through the GM and takes a U-turn to join the internal cerebral vein.[1,2]

In term newborns ICH is relatively uncommon and has a different etiology. ICH in term neonates may be subarachnoid, subdural, intraventricular, parenchymal or epidural in location. In clinical practice, hemorrhage involving multiple compartments is not unusual.[5] Both subdural and subarachnoid hemorrhage in a term newborn is associated with birth trauma either from forceps delivery/vacuum extraction or unassisted vaginal delivery. Vertical moulding of skull causes stretching and tearing of blood vessels of tentorium, falx and dura to produce SDH, while tearing of bridging blood vessels or dural sinuses causes SAH.[5,6] Intraparenchymal hemorrhage (intracerebral or cerebellar) is less frequent than subdural and subarachnoid hemorrhage in term newborns. Intraparenchymal bleeding in the full term newborn may occur as a result of birth asphyxia, instrumental delivery, infection, primary clotting abnormality or congenital vascular abnormality.[7] The incidence of intraventricular hemorrhage (IVH) in term newborns (4.6%) compared to preterm newborn (50%) is very low, probably due to greater maturity of the brain at term. In the term newborn, intraventricular hemorrhage commonly originate from choroid plexus (cryptic hemangioma) or as an extension of the thalamic hemorrhage or subependymal GM bleed.[8] Epidural hemorrhage is rare in term newborn because the middle meningeal artery, which is not yet encased within the bone, moves freely away from displacements of the skull. Epidural hemorrhage, however, may occur in the newborn, in absence of skull fracture, during a difficult forceps extraction when an external forceps causes the outer layer of dura to detach from the inner table of skull. [9]

Several risk factors have been reported in term newborns with ICH. Maternal risk factors causing ICH in the first neonatal week includes usage of drugs (such as aspirin, cocaine), pregnancy-induced hypertension, placental abruption, placental alloimmunization and autoimmune disorders. Major perinatal risk factor are birth trauma, low Apgar score, resuscitation at birth, thrombocytopenia, breastfed infants who received no vitamin K, inherited coagulopathy, disseminated intravascular coagulopathy, increased cerebral venous pressure, prolonged labor, unassisted vaginal delivery, forceps delivery, suction cup, and cesarean-section (sometimes).[10]

3. Clinical presentation

GM-IVH in premature infant is typically diagnosed during the first week of life, 50% on the first day and 90% within the first 4 days. GM-IVH is usually subependymal and asymptomatic, diagnosed by routine screening cranial ultrasound (CUS) in 25-50% of premature infants less than 1,500 gm birth weight and less than 32 weeks' gestation. Clinically symptomatic cases with large hemorrhage and its complication may present with various degree of altered consciousness, cardiorespiratory deterioration, unexplained drop in hematocrit, acidosis, blood glucose alteration, inappropriate antidiuretic hormone secretion, severe apnea or neonatal seizure, bulging fontanelles, abnormal eye movement or alignment, abnormal pupillary response, and abnormal neuromotor examination (hypotonia, decresased motility, tight popliteal angle). [1,3]

The pattern of hemorrhage also differs from GM hemorrhage common in preterm newborns in having a later onset between the 4th and 10th days after birth. Neurologic manifestations like neonatal seizure, decreased level of consciousness, increased intracranial pressure are the most common presentations of ICH in term newborns. The newborn's history, maternal and family history and perinatal risk factors may suggest the diagnosis of ICH.[5]

4. Role of neuroimaging

4.1. Germinal Matrix-Intraventricular Hemorrhage (GM-IVH) in preterm infants

For years neonatal cranial ultrasound (CUS) has been the key diagnostic tool for GM-IVH in premature infants due to its widespread availability, relative low cost, direct bedside scanning and high resolution to detect GM-IVH. CUS is nearly as accurate as CT but much less stressful for an ill premature infant. Diagnostic screening with CUS is recommended in all premature infants (less than 1,500 gm birth weight and less than 32 weeks' gestation) during the second week of life (after which further bleeding is uncommon) or earlier if clinical conditions indicate.[11] Doppler ultrasound can also be used for the imaging and flow velocity measurements of the terminal vein.[12] CT had been used in the original studies to grade the GM-IVH, however, it is no longer recommended for diagnostic purpose due to the adverse effect produce by ionizing radiation on immature brain. MRI is superior to CUS and CT for detection of associated white matter (WM) abnormalities and for identifying hemorrhages particularly small petechial hemorrhage, subacute to chronic hemorrhage and for extracerebral or posterior fossa hemorrhage.[13] The severity of GM-IVH has been evaluated by Papile[14] and Volpe's[15] grading system. When bleeding is confined to the subependymal region, it is classified as grade I; grade II is extension of hemorrhage into non distended lateral ventricle(s) where blood fills less than 50% of the ventricular diameter; in grade III, extensive intraventricular hemorrhage fills greater than 50% of the ventricular diameter leading to hydrocephalus; whereas, grade IV is periventricular hemorrhagic infarction (PVHI).[14,15] Most cases have grade I or grade II intraventricular hemorrhage and do not show late sequelae.[16,17] Infants with grade III or IV hemorrhage have high incidence of neurologic sequelae (approx. 30% will have severe cerebral palsy or mental retardation).[18,19]

Once GM-IVH is diagnosed, follow-up CUS examinations are necessary to determine its complications like periventricular hemorrhagic infarction (PVHI) and post-hemorrhagic hydrocephalus (PHH) and associated cerebellar hemorrhagic injury (CHI), periventricular leukomalacia (PVL), and SAH.[1,4]

Periventricular hemorrhagic infarction (PVHI), classified as grade IV GM-IVH, is a complication of GM-IVH. As previously thought, it is not due to rupture of ependymal lining and extension of intraventricular hemorrhage into the periventricular white matter; rather it occurs due to compression of the terminal vein by GM-IVH resulting in impaired venous drainage and congestion of the medullary veins, which in turn leads to hypoxia-ischemia, infarction and finally hemorrhagic transformation in the periventricular white matter.[20] PVHI can be associated with all grades of GM-IVH (grade I-III) and may be unilateral (65-75% cases) or bilateral (symmetrical or asymmetrical). The severity of PVHI can be graded based on three CUS parameters: (i) extent, (ii) bilaterality, and (ii) the presence of a midline shift. [21] Decreased flow velocity and displacement of the terminal vein can be seen in surviving infants with PVHI using Doppler flow velocimetry.[22] PVHI is in the distribution of the fan-shaped periventricular medullary veins. Intravascular thrombi within the medullary veins can be demonstrated on T2-weighted MRI as linear abnormalities in the WM of centrum semiovale.[20] In living infants, venous infarction usually evolves to produce a porencephalic cyst and rarely multiple cysts. The evolving cyst(s) are associated with the destruction of motor and associative WM axons and preoligodendrocytes. These infants have a high incidence of hemiplegia later in their life.[23] Asymmetric myelination of the posterior limb of the internal capsule on MRI at term has been suggested as an early predictor of hemiplegia in PVHI. [24]

Incidence of ventricular dilatation increases with the severity of GM-IVH. CUS is an ideal tool for monitoring ventricular dilatation in newborns with open fontanelle. Detailed imaging with MRI is usually required prior to shunt surgery and for monitoring the progression or shunt complication after closure of the fontanelle. Posthemorrhagic hydrocephalus (PHH) may be progressive (due to obstruction by blood clot or secondary inflammatory changes which interferes with CSF flow), nonprogressive (due to parenchymal loss secondary to PVHI or PVL) or a combination of the two. PHH survivors with grade III and IV GM-IVH have high risk of significant neurodevelopmental sequelae (e.g. quadriparetic cerebral palsy and/or profound mental retardation).[3,25]

A very high association (77%) has been reported between the cerebellar hemorrhagic injury (CHI) and supratentorial GM-IVH. CHI can result from cerebellar GM hemorrhage (subependymal or subpial), primary hemorrhage, ischemic hemorrhagic transformation of either arterial or venous origin, or dissection of blood through the fourth ventricle or subarachnoid spaces following massive supratentorial GM-IVH. The location of CHI could be unilateral, vermian, bilateral or a combination of these. Mastoid CUS view helps to detect CHI in 3% of premature infants. MRI detects small petechial cerebellar hemorrhages not visible on CUS. CHI may eventually lead to several types of atrophic changes. 40% of CHI survivors have global developmental (cognitive and social communication disabilities) and functional deficits (motor deficits).[3,26]

A strong association between GM-IVH and periventricular leukomalacia (PVL) has been suggested. GM-IVH and PVL may develop in parallel, the ischemic injury may injure the GM and the periventricular WM leading to both GM-IVH and PVL. The CUS markers of cystic and diffuse PVL include echolucencies, echodensities and nonprogressive ventriculomegaly.[3,27]

Subarachnoid hemorrhage (SAH) is relatively common in premature infants with GM-IVH. The true incidence of SAH is not known as it is difficult to visualize the extra-axial hemorrhage by CUS. MRI is the modality of choice to detect SAH. SAH could be one of the reasons for PHH (obstructive arachnoiditis), neonatal seizure (irritation of cerebral convexity) and cerebral/cerebellar growth impairment.[28,29]

Impaired cerebellar and supratentorial gray matter growth has been reported in GM-IVH survivors. The complicated GM-IVH (PVHI and PVL) causes impaired growth and development of the contralateral cerebellar hemisphere due to injury to specific supratentorial projections (crossed cerebellar diaschisis).[28] The uncomplicated GM-IVH (without parenchymal involvement) is associated with impaired growth of the supratentorial gray matter, probably because the GM destruction prevents the neuronal and astrocyte precursor cells from reaching their cortical destination. In addition, SAH (circulating free radical) may directly injure the cerebral cortex surface.[29]

4.2. Intracranial hemorrhage (ICH) in term newborns

Subarachnoid hemorrhage (SAH) is the most common type of hemorrhage among the symptomatic term newborns.[6] On FLAIR (fluid attenuated inversion recovery) sequence of MRI, SAH is detected as hyperattenuating fluid in the basal subarachnoid spaces or along the cerebral sulci. Extensive subarachnoid hemorrhage may be difficult to distinguish from subdural hemorrhage and the two may co-exist. [3]

Subdural hemorrhage (SDH) is the most frequent hemorrhage among asymptomatic term newborns.[5] SDH is usually infratentorial, may result from rupture of the vein of Galen, straight or transverse sinus. On imaging, SDH can be seen as a crescent-shaped hyperattenuating region conforming to the adjacent brain. Small posterior fossa SDHs are common in infants and are usually of no clinical significance; however, when large, may result in compression of the brain stem and death. Infratentorial SDH may be difficult to distinguish from transverse sinus thrombosis. The two may co-exist or a SDH may compress the sinus predisposing to thrombosis. Convexity SDH are less common than posterior fossa hemorrhage and the two may co-exist. Rupture of superficial cortical veins gives rise to convexity SDH, which may be accompanied by SAH. Convexity SDH is mainly unilateral. Large convexity hemorrhage may be associated with infarction of the brain either from arterial occlusion or impaired venous drainage. Associated parenchymal hemorrhage either due to hemorrhagic tendency or in association with infarction may also occur.[3]

Large SDH may result in impairment of CSF flow and associated ventricular dilatation or widening of the extracerebral space (external hydrocephalus). The evolution of SDH may result in the formation of a subdural effusion which may remain at the site of a previous SDH for months and may be associated with rebleeding.[3]

Parenchymal hemorrhagic lesions may co-exist with hemorrhage elsewhere in the cranium. Parenchymal hemorrhage may be focal or multifocal and of any size. Multifocal small hemorrhages may be found in term infants presenting with convulsions during the first few days of life. Thalamic hemorrhage is usually unilateral and associated with IVH. Primary thalamic hemorrhage needs to be distinguished from the bilateral thalamic abnormalities seen in HIE. The thalamic lesions in HIE are focal, usually involving the lateral thalamic nuclei and sometimes the medial nuclei, these lesion have high signal intensity on T1-weighted images and low signal intensity on T2-weighted images due to capillary proliferation in region of infarction (not due to hemorrhage). However, infants with HIE may develop large intracranial hemorrhage. Basal ganglia hemorrhage may occur as an isolated event in term newborn, sometimes it is difficult to differentiate it from a hemorrhagic infarction involving a deep branch of middle cerebral artery. Cerebellar hemorrhage may be primary, secondary to venous infarction or may complicate massive intraventricular or subarachnoid hemorrhage. The MR appearance of parenchymal hemorrhage varies with time depending on the oxidation state of hemoglobin.[30]

5. Prevention and management

5.1. GM-IVH in preterm infants

Modern practice should emphasize on (i) prevention of GM-IVH, (ii) halting its progression, and (iii) reducing its complications. Important preventive measures include (i) special obstetric care for high-risk pregnancies, (ii) treatment of bacterial vaginosis for reducing premature delivery, (iii) prevention of imminent premature labor using tocolytic agents, cesarean section delivery in selected cases, and (iv) maternal administration of magnesium sulfate. In addition, optimal ventilation and strict hemodynamic control of the premature infant are the cornerstones of preventing GM-IVH and its progression. Several postnatal pharmacological agents for prevention of GM-IVH are still under trial.[25,31,32]

PHH, a complication of GM-IVH has a very unpredictable course. 60% of the infants may undergo spontaneous resolution while 40% may require a ventriculoperitoneal shunt, the definitive treatment of progressive PHH.[25] Repeated lumbar puncture may temporarily arrest the progression of PHH, but the long-term benefit of this approach remains unknown.[3]

5.2. ICH in term newborns

The single most important primary prevention is successful accomplishment of a vaginal delivery with or without obstetric instrument. However, forceful vaginal delivery should not be attempted if either vacuum extraction or forceps delivery has failed.[33]

Medical interventions should be implemented on the very first clinical suspension of ICH. The goal of medical therapy is to provide adequate ventilation, prevent metabolic acidosis, to keep vital organs well perfused and to control seizure activity. Sick newborns with ICH are managed in the intensive care unit. Any treatable etiological factor (e.g. sepsis, dehydration, thrombocytopenia, vitamin K deficiency or coagulopathy) should be identified and treated promptly. Most symptomatic newborns with intracranial hemorrhage do not require neurosurgical intervention. Neurosurgical intervention could, however, be lifesaving in a situation in which there is a sudden clinical deterioration primarily due to rise in intracranial pressure as a result of massive ICH and PHH.[34] The secondary prevention is to limit the extent of parenchymal injury to the brain due to neurosurgery or hematoma.[35]

6. Summary

GM-IVH and its complications have potential impact on morbidity, mortality and long-term neurodevelopmental outcome. The mechanism of GM-IVH is multifactorial and involves a combination of vascular anatomic immaturity and complex hemodynamic factors. Goals are prevention of GM-IVH, halting its progression and reducing its complications.

A strong association between traumatic deliveries especially vacuum extraction/forceps delivery and hemorrhagic lesions has been well established in term newborns. Hemorrhage is often present at more than one site. MRI can be used to time the onset of lesions. Hemorrhage may be primary or secondary and occur within an arterial or venous infarct. Neurodevelopmental outcome varies with the site of hemorrhage and the underlying cause. In the majority of term newborns with ICH, medical therapy is the primary mode of treatment; rarely, surgical intervention may be needed in selected cases.

Author details

Shahina Bano[1*], Vikas Chaudhary[2], Umesh Chandra Garga[1], Sachchidanand Yadav[1] and Sachin Kumar Singh[3]

*Address all correspondence to: dr_shahinaindia@yahoo.com

1 Department of Radiodiagnosis, PGIMER and Dr. Ram Manohar Lohia (RML) Hospital, India

2 Department of Radiodiagnosis, Employees' State Insurance Corporation (ESIC) Model Hospital, Gurgaon, Haryana, India

3 Department of Radiodiagnosis, Jawahar Lal Nehru (JLN) Medical College, Bhagalpur, Bihar, India

References

[1] Volpe JJ. Intracranial hemorrhage: neurology of the newborn. 5th edition. Philadelphia: W.B. Saunders Co.; 2008. P. 481-588.

[2] Pape KE, Wigglesworth JS. Hemorrhage, ischemia and the perinatal brain. Philadelphia: J.B. Lippincott;1979.

[3] Haim Bassan. Intracranial hemorrhage in the preterm infant: understanding it, preventing it. Clin Perinatol 2009;36:737-62.

[4] Du Plessis AJ. Cerebrovascular injury in premature infants: current understanding and challenges for future prevention. Clin Perinatol 2008;35:609-41.

[5] Looney CB, Smith JK, Merck LH, et al. Intracranial hemorrhage in asymptomatic neonates: prevalence on MR images and relationship to obstetric and neonatal risk factors. Radiology 2007;242:535-41.

[6] Fenichel GM, Webster DL, Wong WK. Intracranial hemorrhage in the term newborn. Arch Neurol 1984;41:30-4.

http://dx.doi.org/10.5772/58476

[7] Sandberg DI, Lamberti-Pasculli M, Drake JM, Humphreys RP, Rutka JT. Spontaneous intraparenchymal hemorrhage in full-term neonates. Neurosurgery 2001;48:1042-8.

[8] Hayden CK Jr. Shattuck KE, Richardson CJ, Ahrendt DK, House R, Swischuk LE. Subependymal hemorrhage in full-term neonates. Pediatrics 1985;75:714-8.

[9] Volpe JJ. Intracranial hemorrhage: subdural, primary subarachnoid, intracerebellar, intraventricular (term infant), and miscellaneous. In: Neurology of the newborn. 4th ed. Philadelphia: WB Saunders, 2001:397-423.

[10] Jhawar BS, Ranger A, Steven D, Del Maestro RF. Risk factors for intracranial hemorrhage among full-term infants: a case-control study. Neurosurgery 2003;52:581-90.

[11] Baun J. Neonatal intracranial hemorrhage. Journal of Diagnostic Medical Sonography 1991;7:121-31.

[12] Quinn M, Ando Y, Levene M. Cerebral arterial and venous flow-velocity measurements in post-hemorrhagic ventricular dilation and hydrocephalus. Dev Med Child Neurol 1992;34:863-9.

[13] Maalouf EF, Duggan PJ, Counsell SJ, et al. Comparison of findings on cranial ultrasound and magnetic resonance imaging in preterm infants. Pediatrics 2001;107:719-927.

[14] Papile LA, Burstein J, Burstein R, et al. Incidence and evolution of subependymal and intraventricular hemorrhage: a study of infants with birth weights less than 1,500 gm. J Pediatr 1978;92:529-34.

[15] Volpe JJ. Intracranial hemorrhage: germinal matrix-intraventricular hemorrhage of the premature infant. In: Neurology of the newborn. Philadelphia: W.B. Saunders; 1995. p. 403-62.

[16] Kuban KC, Allred EN, O'Shea TM, et al. Cranial ultrasound lesions in the NICU predict cerebral palsy at age 2 years in children born at extremely low gestational age. J Child Neurol 2009;24:63-72.

[17] Laptook AR, O'Shea TM, Shankaran S, et al. Adverse neurodevelopmental outcomes among extremely low birth weight infants with a normal head ultrasound: prevalence and antecedents. Pediatrics 2005;115:673-80.

[18] Bassan H, Limperopoulos C, Visconti K, et al. Neurodevelopmental outcome in survivors of periventricular hemorrhagic infarction. Pediatrics 2007;120:785-92.

[19] Broitman E, Ambalavanan N, Higgins RD, et al. Clinical data predict neurodevelopmental outcome better than head ultrasound in extremely low birth weight infants. J Pediatr 2007;151:500-5,505.e1-2.

[20] Counsell SJ, Maalouf EF, Rutherford MA, et al. Periventricular haemorrhagic infarct in a preterm neonate. Europ J Paediatr Neurol 1999;3:25-7.

[21] Bassan H, Benson CB, Limperopoulos C, et al. Ultrasonographic features and severity scoring of periventricular hemorrhagic infarction in relation to risk factors and outcome. Pediatrics 2006;117:2111-8.

[22] Taylor GA. Effect of germinal matrix hemorrhage on terminal vein position and patency. Pediatr Radiol 1995;25:1537-40.

[23] Volpe JJ. Brain injury in premature infants: a complex amalgam of destructive and developmental disturbances. Lancet Neurol 2009;8:110-24.

[24] de Vries LS, Groenendaal F, van Haastert IC, et al. Asymmetrical myelination of the posterior limb of the internal capsule in infants with periventricular haemorrhagic infarction: an early predictor of hemiplegia. Neuropediatrics 1999;30:314–9.

[25] Murphy BP, Inder TE, Rooks V, et al. Posthaemorrhagic ventricular dilatation in the premature infant: natural history and predictors of outcome. Arch Dis Child Fetal Neonatal Ed 2002;87:F37-41.

[26] Limperopoulos C, Benson CB, Bassan H, et al. Cerebellar hemorrhage in the preterm infant: ultrasonographic findings and risk factors. Pediatrics 2005;116:717-24.

[27] de Vries LS, Van Haastert IL, Rademaker KJ, et al. Ultrasound abnormalities preceding cerebral palsy in high-risk preterm infants. J Pediatr 2004;144:815-20.

[28] Limperopoulos C, Soul JS, Haidar H, et al. Impaired trophic interactions between the cerebellum and the cerebrum among preterm infants. Pediatrics 2005;116:844-50.

[29] Vasileiadis GT, Gelman N, Han VK, et al. Uncomplicated intraventricular hemorrhage is followed by reduced cortical volume at near-term age. Pediatrics 2004;114:e367-72.

[30] Zuerrer M, Martin E, Boltshauser E. MR imaging of intracranial hemorrhage in neonates and infants at 2.35 Tesla. Neuroradiology 1991;33:223-9.

[31] Leviton A, Fenton T, Kuban KC, et al. Labor and delivery characteristics and the risk of germinal matrix hemorrhage in low birth weight infants. J Child Neurol 1991;6:35-40.

[32] Di Renzo GC, Mignosa M, Gerli S, et al. The combined maternal administration of magnesium sulfate and aminophylline reduces intraventricular hemorrhage in very preterm neonates. Am J Obstet Gynecol 2005;192:433-8.

[33] Benedetti TJ. Birth injury and method of delivery. N Engl J Med 1999;341:1758-9

[34] Hanigan WC, Powell FC, Miller TC, Wright RM. Symptomatic intracranial hemorrhage in full-term infants. Childs Nerv Syst 1995;11:698-707.

[35] Tekgul H, Gauvreau K, Soul J, et al. The current etiologic profile and neurodevelopmental outcome of seizures in term newborn infants. Pediatrics 2006;117:1270-80.

Chapter 2

Gender Differences in Incidence, Pathophysiology, and Outcome of Primary Intracerebral Hemorrhage

Michael James and Sankalp Gokhale

Additional information is available at the end of the chapter

http://dx.doi.org/10.5772/58469

1. Introduction

While ongoing clinical trials are directly assessing ethnic/racial differences in incidence and outcome of ICH, it is clear that individuals of certain ethnic groups (Asian, African American, etc.) more commonly experience ICH [1,2]. However, few published studies comprehensively assess the role of gender differences in incidence, clinical presentation, etiology, or outcome after ICH. This knowledge is important for understanding putative pathophysiological mechanisms, improving research models, and developing effective treatment options for patients with ICH.

2. Incidence

While epidemiological observations on gender differences in the incidence of ischemic stroke are abundant [3], relatively few have been published assessing gender differences in incidence of ICH [4,5]. Over the course of the last several years, the incidence of primary ICH in the developed world seems to be same or decreasing [4,5]. Presumably this is attributed to better public awareness of preventive measures, such as control of hypertension, regular physical activity, and healthy lifestyle. This observation is supported by population-based studies by Islam et al. [6] in Perth, Australia as well by Lovelock et al. [5] in Oxfordshire, UK. On the other hand, the incidence of ICH was reported to be nearly the same in studies done in France and Finland [7,8]. In a systematic meta-analysis involving 36 studies over last 2 decades, Charlotte et al. [4], found no significant difference in incidence of ICH over time.

3. Influence of gender on incidence of ICH

Observations differ regarding the existence of gender differences in incidence of ICH. Further, ethnic/racial background and age may interact with gender to influence ICH incidence. For example, gender differences in ICH incidence are difficult to find in populations composed largely of Caucasian individuals, such as Europe and Australia. In contrast, data from multiethnic/racial populations, such as Northern America, suggest gender differences do exist. Interestingly, there is a relative paucity of data from African, Southeast Asian, and Middle Eastern populations about gender differences in incidence of ICH.

Below observations from major stroke registries and epidemiological studies are summarized in Table 1. Statistical significance is reported wherever available (p values and confidence intervals). All the incidence values for ICH are reported as per 100,000 people per year, unless otherwise stated.

3.1. Studies from Europe

In a community-based study in Oxfordshire, UK, Bamford et al. [9] reported no gender difference in overall incidence of ICH for men compared to women. Out of 66 patients with ICH presenting over 5 years (1981-1986), 29 were men and 37 were women. Out of 37 women with ICH, 23 (62%) were more than 75 years old, as compared to only 6 out of 29 men (21%). In a population-based observational study during the years 1985-1989 in Dijon, France, involving a town with 140,000 subjects, no gender differences were found in 87 cases of ICH from a total of 984 cases of new strokes [10].

Fogelholm et al. [11] studied the epidemiology of ICH in central Finland during the years 1985-1989. From a total of 158 patients with ICH, they noted that 80 were women and 78 were men. The incidence rates were similar in men (32) as compared to women (31) over this time period. There was no difference in age of presentation between men and women with ICH.

These results were similar in a community-based study over the year 1989 in Italy. Researchers in this study did not find any gender difference in annual incidence of ICH [12].

In a similar population-based study in Italy in 1992, it was noted that incidence of ICH was similar in men and in women [13]. The incidence was 47 (95% CI=35-61) in women as compared to 39 (95% CI=27-53) in men. In addition, there was no significant difference in the age of presentation with the majority of patients being 65-75 years old at the time of presentation.

In the FINSTROKE study, Sivenius et al. [8] looked at the incidence of various types of strokes across ages and genders from 1983-1997. This was a large comprehensive study to evaluate incidence and types of various strokes, in addition to outcomes in Finland, over a span of 15 years. The investigators evaluated 5650 new cases of stroke. In their study, the incidence of ICH did not change as opposed to incidence of ischemic stroke, which showed a decreasing trend. Further the incidence of ICH was similar in men and women during course of the study ($p=0.6$).

The incidence of ICH was similar between men and women in an observational study conducted at Vibo Valentina, Italy. This was a population-based study involving 97 subjects with primary ICH in southern Italy conducted in 1996. The investigators found that the incidence of ICH was 35. The incidence ratio in women to men was 0.92 (95% CI=0.56-1.51). This was similar to another observation in 1989 from Valle d'Aosta, Italy. The investigators noted an annual incidence of primary ICH of 22 in men and 30 in women (95% CI=0.85-3.49) [12,14,15].

These findings are similar to Arcadia Stroke Study from Greece. Vemmos et al. [16] studied 555 patients with stroke during 1993-1995 from a population-based stroke registry database study. The age-adjusted incidence was 50.9, similar to other European studies. The incidence of ICH was 64.4 (95% CI=41–88) in men as compared 38.2 (95% CI=20–56) in women. Thus, there was no gender difference in incidence of ICH in this cohort.

The incidence of ICH over a 2-year period was found to be 24 in the German study, ESPro [17]. This was a prospective population-based study to evaluate the incidence of various types of strokes. The incidence of ICH was 29 in women and 18 in men.

3.2. Studies from Asia

In the population-based Hisayama study from Japan, Kubo et al. [18] analyzed patients with various types of strokes from 1961-2000, breaking down patient populations into 3 time-based cohorts (1961-1973, 1974-1986, 1986-2000). They observed differences in the incidence of ICH in men as compared to women. Incidence in men was higher (321, 125, and 130 over the 3 time periods) as compared to women (63, 73, and 70, respectively) over the course of these cohorts. Notably, there was a drop in incidence of ICH in men from 321 to 125 between the first and second time periods (p=0.01), whereas the incidence in women remained the same across 40 years of observation.

In a large population-based retrospective observational study involving 32,859 inhabitants of rural Japan, Morikawa et al. [19] found the incidence of ICH was lower in women as compared to men. The researchers analyzed the incidence of various types of strokes over the years 1977-1991, divided into 3 time periods (1977-1981, 1982-1986 and 1986-1991). There were 410 cases of ICH [175 (97 men, 78 women), 120 (70 men, 50 women) and 115 (67 men, 48 women) during three sub-periods]. The incidence of ICH in each time period for men was 605, 455, and 417 as compared to 476, 322, and 329 for women. Further, in their study they did not notice any difference in the age of presentation between the genders.

In another hospital-based study by Inagawa et al. [20], the incidence of ICH was similar in men and women. They studied 350 patients with first-ever primary ICH who were treated during the 8-year period 1991 to 1998 in Izumo City, Japan. The incidence of ICH in men was 93 as compared to 57 in women. (p=0.938). Inagawa et al carried out a population-based retrospective study of 267 patients with primary ICH in Izumo City, Japan, during the time period 1991-1996. The investigators noted similar incidence of ICH in men and women. The incidence of ICH in age groups 50-59, 60-69, 70-79 and 80-89 in men versus women was: (36/13, 55/27, 40/41, 20/33). The study did not have enough power to assess for statistical significance. The slightly higher incidence of ICH in older women was attributable to higher life expectancy in women and increasing incidence of amyloid angiopathy with age. In another population-based study in Shibata province of Japan, the investigators studied 97 patients with primary ICH during the time period 1976-1978. The investigators noted incidence in men of 80 as compared to 44 in women. [18,19,21-23].

In a large multicenter study conducted across 10 different provinces in China during the years 1991-2000, the investigators identified a total of 16,031 cases of new onset stroke. Out of these cases, 23.9 percent were ICH. The investigators found no difference in incidence of ICH across gender. [24]

3.3. Studies from North America

In the Northern Manhattan Stroke study [25,26] the investigators found that the annual incidence of ICH in the urban population of New York City was 30.9. Incidence was the same in men (81) as compared to 74 in women. Risk of ICH in men was higher than in women overall with a relative risk of 1.5 (95% CI=1.2-1.8) for ICH in men as compared to women. Below age 65, men were at significantly greater risk than women (RR=3.4, 95% CI=2.7-4.3) but age 65 and above men and women had a similar risk (RR=0.8, 95% CI=0.5-1.2). Men were at significantly greater risk of deep ICH than women (RR=1.8, 95% CI=1.4–2.3) but there was no difference in risk of lobar ICH (RR=1, 95% CI=0.8-1.2).

Kissela et al. [27] conducted a hospital-based observational study of 3136 stroke patients during the time period of January 1, 1993, to June 30, 1994, to evaluate racial differences in clinical features of strokes. The investigators noted that the incidence of primary ICH was 37 (95 CI, 28–46) in blacks as compared to 18 (95 CI, 16–20) in whites. Further, the incidence of ICH was higher in black men in age group 65-74, (225) as compared to black women in the same age group (100). There was no significant gender difference in the incidence of ICH in white population.

3.4. Studies from South America

There have been few studies from South America in this regard. The PISCIS stroke project, a community-based prospective study in Iquique, Chile noted that the incidence of ICH was higher in men as compared to women with nearly two thirds of the total study cohort being men. Minelli et al. [28] in a population-based study in Brazil noted similar findings. They studied overall incidence of various types of strokes in the assigned population and found the overall incidence of ICH was 14.7. Incidence in men (18.7) was nearly twice as much in women (10.7). The majority of women were older (above 75 years of age) at presentation as compared to men who most commonly were between 55-74 years of age.

3.5. Studies from Australia and New Zealand

In STROMA study, slightly higher incidence of ICH in men was found as compared to women [29]. These findings are similar to the Australian study, NEMESIS, completed over period of 2 years [30]. A higher incidence of ICH was observed in men compared to women (30 vs. 18) [30,31].

There was no significant gender difference in incidence of ICH in Auckland Regional Community Stroke Study [32-34]. The adjusted incidence for ICH was 29, with an incidence of 33 in men and 26 in women. There was uniform increase in the incidence of ICH in the older age groups in both genders.

4. Gender differences in pathophysiology and outcome

While gender differences in clinical features of aneurismal subarachnoid hemorrhage has been studied [35,36], a relative paucity of data exists regarding gender difference in similar characteristics after ICH.

In the Northern Manhattan study [25,26], the incidence of ICH in men was 81 as compared to 74 in women (RR=1.5, 95% CI=1.2-1.8). The incidence of deep ICH was 58 in men as compared to 43 in women (RR=1.8, 95% CI=1.4–2.3). On the other hand, the incidence of lobar ICH was 21 in men as compared to 28 in women (RR=1, 95% CI 0.7–1.4).

Inagawa et al. [20] evaluated incidence of ICH across gender and anatomical location during 8-year period 1991 to 1998 in Izumo City, Japan. Though ICH occurred more commonly in men as opposed to women, the investigators did not find any significant difference regards to the anatomical hematoma location across genders. Taken together, these findings suggest a potential interaction between gender and race/ethnicity in determining the site of ICH.

Mean arterial pressure (MAP) may be associated with mortality in first 24 hours after presentation of ICH. Qureshi et al. [37] carried out a retrospective chart review of 105 patients with a diagnosis of primary ICH, and looked at the effect of rapid lowering of MAP on mortality in the first 24 hours after ICH. The rate of decline in MAP (slope) was independently associated with increased mortality (p=0.04), i.e., a faster rate of MAP decline was associated with higher mortality. This effect was independent of other known predictors of mortality, i.e., hematoma volume, presence of ventricular blood, and initial GCS score. While not powered to examine gender-specific interaction, MAP lowering effects on mortality were nearly greater in men as compared to women. (p=0.08). These findings may highlight gender differences in susceptibility to changes in MAP and its potential contribution to mortality.

Gender differences may exist in formation of deep venous thrombosis (DVT) in patients with primary ICH. Kawase et al. [38] prospectively evaluated 81 patients with primary ICH for risk of developing DVT. After adjustment for age and relevant confounders, female sex was the only independent predictor for DVT (odds ratio 6.89, 95% confidence interval, CI, 1.56-36.34, p=0.014). Female patients with an initial NIHSS score ≥ 12 had 19 times the risk for DVT compared to men with an NIHSS score <12 (95% CI 2.61-213.77, p=0.007). Development of DVT could add significantly to the morality and morbidity, especially in patients with ICH, as these patients are not routine candiates for anticoagulation in the acute phase.

Perihematomal edema (PHE) formation after ICH may affect mortality and long-term neurological function in this patient population. Wagner et al. [39] looked at the role of gender in perihematomal edema in cases of primary ICH. PHE development was assessed over a 14-day period on follow-up CT scans in 387 subjects and was compared between men and women. The investigators found that starting at days 2-4, women showed lower PHE values ($P<0.05$; days 2-4 and 8-11). The mechanism and ultimate effect on outcome of these findings is unclear.

To further support the neuroprotective role of female sex hormones, menopause may alter the risk of primary ICH in women. Feldman et al. [40] carried out a prospective observational study of 1714 patients with a diagnosis of ICH, and evaluated various risk factors for developing primary ICH. The investigators found that post-menopausal women had signifcantly higer incidence of ICH as compared to premenopausal women. Thus, menopause was significantly associated with development of primary ICH (adjusted OR, 2.50; 95% CI, 1.06 to 5.88). This observation may support role of female sex hormones in modifying the risk of developing ICH. Clearly, more studies are necessary in this regard.

Five studies provide information about the fatality of ICH in both genders. In a cross sectional study done by Kimura et al. [41], investigators found no difference in ICH mortality in men vs. women at 28 days. The case fatality was 17.6% overall, with 18.8% in men and 16.2% in women. These findings were similar to subsequent studies. Data from the Arcadia stroke registry [16] showed a slightly higher rate of case fatality in women (51.8%) as compared to men (44%) at 4 weeks after ICH. There was no difference in the case fatality rates in men as compared to women from the observations in STROMA study. [29] The investigators found that case fatality rate was 23.9% in men and 22.7% in women at 4 weeks after UCG. These findings were supported by a large multicenter study from China, where the case fatality rate was 48.4 percent in men, similar to 50.7 percent in women at 4 weeks after ICH. [24] Strikingly different were the observations made by Thrift et al. [30] from Australia. The investigators found that case fatality was higher in women (50.6%) as compared to men (29.2%). Finally, women had better survival than men after first-ever primary ICH in a prospective stroke register from Sweden, largely explained by a higher 28-day mortality in male patients over 75 years. [42]

5. Preclinical observations of the role of sex in ICH

Data from preclinical models of ICH suggest response to the estrogen therapy. Nakamura et al. [43] studied sex differences in rat model of ICH. They observed that brain edema and neurological deficits at 24 hours after ICH were less in female rats as compared to male rats. The investigators then studied the role of an estrogen derivative on edema as well as functional outcome in male and female rats. They observed that administration of exogenous estrogen decreased edema as well as neurological deficits in male rats, but made no difference in female rats. Other studies support these findings. [44,45] These observations may also extend to pretreatment. Auriat et al. [46] found that estrogen pretreatment significantly reduced hemorrhagic blood volume at 12 hours after ICH in male rats; however estrogen did not lessen cerebral edema at 2 days after ICH. These observations are supported by work from Gu et al. [47] The investigators evaluated the role of estrogen pretreatment in reducing brain edema and neuronal survival in iron induced injury in murine models of ICH. The investigators found that estrogen pretreatment in male rats reduced brain edema ($p < 0.01$) as well as reduced neuronal death *in vitro* suggesting a broader neuroprotective effect of etrogen.

Further, accumulation of iron and free radicals is proposed as one of the mechanisms of neuronal injury after ICH. Chen et al. [48] evaluated the molecular mechanisms underlying estrogen-mediated neuroprotective effect against iron induced neuronal injury in cases of ICH. The investigators found that ferrous citrate induced greater brain injury in male rats than female rats. Further, they observed that estrogen pretreatment was protective in iron-induced brain injury in both sexes but the effect was more pronounced in female rats as compared to male rats. The protective effect in female rats was attributed to higher concentrations of estrogen receptor alpha in the brain regions involved. These observations suggest a direct receptor mediated neuroprotective effect of estrogens against iron-induced free radical injury in murine models of ICH.

Findings from work with estrogen may be partialy extended to other gonadal hormones. Chen et al. [49] evaluated progesterone and testosterone effects on ICH-induced brain injury in male rats. There was significant reduction of PHE in progesterone-treated rats ($p < 0.05$) as well as improved functional outcome following ICH ($p < 0.05$), as compared to vehicle-treated rats. However, testosterone treatment did not affect PHE and was associated with worse functional outcome ($p < 0.05$) as compared to vehicle-treated rats. These observations suggest a differential effect between female and male sex hormones after ICH.

Finally, Lei et al. [50] evaluated the role of sex and APOE polymorphism in modifying outcomes in murine models of ICH. The investigators found that female mice had better functional outcome as compared to male mice after ICH, as measured by neurobehavioral and cognitive tests. Further, female mice showed rapid rates of recovery as compared to male mice after comparable ICH injury. Interestingly, both male and female mice showed functional benefits after administartion of apoE mimetic peptide. Thus, sex differences may exist in pharmacogenomic interactions for drug therapy to improve recovery after ICH.

6. Conclusions

Several general conclusions can be made from the above observations. Gender differences in incidence of ICH appear to exist in Asian and South American populations, where men seem to suffer a higher incidence. Gender differences in incidence of ICH are not as obvious in Northern America, European, or Australian-New Zealand populations. Interestingly, there appears to be an interaction between age and gender in many of these populations, with ICH occurring at a younger age in men. Differences in access to preventative treatment, prevalence of known risk factors (e.g., hypertension, alcohol consumption, use of sympathomimetic drugs), and genetic variance may explain some of these disparities.

Less clear is the existence of gender differences in pathophysiology and outcomes after ICH. While preclinical data support the role of gonadal hormones influencing hemostatic and neuroinflammatory modulation after ICH, their effects on recovery in humans are unknown.

No.	Author	Study details	Year/s	Observations
1	Tanaka (1981) [23]	Shibata, Japan	1976-1978	Higher incidence of ICH in men
2	Bamford (1988) [9,51]	Oxford shire, UK	1981-86	Same incidence of ICH across genders. Women older age of presentation
3	Giroud (1991) [10]	Dijon, France	1985-1989	No difference in incidence of ICH across genders
4	D'Alessandro (1992) [12]	Italy	1989	No difference in incidence of ICH across genders
5	Fogelholm (1992) [11]	Central Finland	1985-1989	No difference in incidence of ICH across genders, women older age of presentation
6	Lauria G (1995) [13]	Italy	1992	Slightly higher incidence of ICH in women
7	Kimura (1998) [41]	Okinawa, Japan	1988-1991	No difference in ICH related mortality at day 30 across genders
8	Kolominsky-Rabas (1998) [17]	Germany	1993-1995	Slightly higher incidence of ICH in women
9	Sacco (1998) [26]	New York, USA	1993-1996	Higher incidence of ICH in men but women older at age of presentation
10	Vemmos (1999) [16]	Greece	1993-1995	Much higher incidence of ICH in older men
11	Morikawa (2000) [19]	Rural Japan	1977-1991	Same incidence and age of presentation of ICH across genders
12	Thrift (2001) [30]	Australia	1996-1997	Incidence of ICH more in men as compared to women
13	Di Carlo (2003) [14]	Calabria, Italy	1996	Slightly higher incidence of ICH in men
14	Inagawa (2003) [20]	Izumo, Japan	1991-1998	Incidence of ICH more in men as compared to women
15	Kubo (2003) [18]	Japan	1961-2000	Significantly higher incidence of ICH in men
16	Zhang (2003) [24]	China	1991- 2000	No difference in mortality in ICH across genders
17	Sivenius (2004) [8]	Finland	1983-1997	Higher incidence of ICH in men

No.	Author	Study details	Year/s	Observations
18	Anderson (2005) [34]	Auckland, NZ	1981-2003	No difference in incidence of ICH across genders
19	Fogelholm (2005) [52]	Finland	1985-1991	Higher incidence of ICH related mortality in men
20	Khan (2005) [29]	Sweden	1989-1999	Higher incidence of ICH in men
21	Labovitz (2005) [25]	Manhattan, USA	1993-1996	Higher incidence of ICH in men of African American ethnicity
22	Lavados (2005) [53]	Iquique, Chile	2000-2002	Higher incidence of ICH in men
23	Benatru (2006) [7]	Dijon, France	1985 to 2004	Same incidence of ICH over years across genders
24	Feigin (2006) [33]	Auckland, NZ	2002-2003	Same incidence of ICH across genders
25	Lovelock (2007) [5]	Oxford shire, UK	1981-2006	Decrease in incidence of ICH across genders over years
26	Minelli (2007) [28]	Matão, Brazil	2003-2004	Higher incidence of ICH in men
27	Islam (2008) [6]	Perth, Australia	1989-2001	Decrease in incidence of ICH across genders over years
28	Thrift (2009) [31]	Melbourne, Australia	1997-1999	Higher incidence of ICH in men
29	Lavados (2010) [54]	Iquique, Chile	2000-2002	The incidence of non-lobar ICH was high, most non-white populations

Author details

Michael James[1*] and Sankalp Gokhale[2]

*Address all correspondence to: Michael.james@duke.edu

1 Department of Neurology and Anesthesiology, Duke University Hospital, Duke University School of Medicine, NC, USA

2 Department of Neurology, Duke University Hospital, Duke University School of Medicine, NC, USA

References

[1] Flaherty ML, Woo D, Haverbusch M, et al. Racial variations in location and risk of intracerebral hemorrhage. *Stroke; a journal of cerebral circulation.* May 2005;36(5): 934-937.

[2] Mayer SA, Rincon F. Treatment of intracerebral haemorrhage. *Lancet neurology.* Oct 2005;4(10):662-672.

[3] Feigin VL, Lawes CM, Bennett DA, Barker-Collo SL, Parag V. Worldwide stroke incidence and early case fatality reported in 56 population-based studies: a systematic review. *Lancet neurology.* Apr 2009;8(4):355-369.

[4] van Asch CJ, Luitse MJ, Rinkel GJ, van der Tweel I, Algra A, Klijn CJ. Incidence, case fatality, and functional outcome of intracerebral haemorrhage over time, according to age, sex, and ethnic origin: a systematic review and meta-analysis. *Lancet neurology.* Feb 2010;9(2):167-176.

[5] Lovelock CE, Molyneux AJ, Rothwell PM, Oxford Vascular S. Change in incidence and aetiology of intracerebral haemorrhage in Oxfordshire, UK, between 1981 and 2006: a population-based study. *Lancet neurology.* Jun 2007;6(6):487-493.

[6] Islam MS, Anderson CS, Hankey GJ, et al. Trends in incidence and outcome of stroke in Perth, Western Australia during 1989 to 2001: the Perth Community Stroke Study. *Stroke; a journal of cerebral circulation.* Mar 2008;39(3):776-782.

[7] Benatru I, Rouaud O, Durier J, et al. Stable stroke incidence rates but improved case-fatality in Dijon, France, from 1985 to 2004. *Stroke; a journal of cerebral circulation.* Jul 2006;37(7):1674-1679.

[8] Sivenius J, Tuomilehto J, Immonen-Raiha P, et al. Continuous 15-year decrease in incidence and mortality of stroke in Finland: the FINSTROKE study. *Stroke; a journal of cerebral circulation.* Feb 2004;35(2):420-425.

[9] Bamford J, Sandercock P, Dennis M, Burn J, Warlow C. A prospective study of acute cerebrovascular disease in the community: the Oxfordshire Community Stroke Project--1981-86. 2. Incidence, case fatality rates and overall outcome at one year of cerebral infarction, primary intracerebral and subarachnoid haemorrhage. *Journal of neurology, neurosurgery, and psychiatry.* Jan 1990;53(1):16-22.

[10] Giroud M, Milan C, Beuriat P, et al. Incidence and survival rates during a two-year period of intracerebral and subarachnoid haemorrhages, cortical infarcts, lacunes and transient ischaemic attacks. The Stroke Registry of Dijon: 1985-1989. *International journal of epidemiology.* Dec 1991;20(4):892-899.

[11] Fogelholm R, Nuutila M, Vuorela AL. Primary intracerebral haemorrhage in the Jyvaskyla region, central Finland, 1985-89: incidence, case fatality rate, and functional outcome. *Journal of neurology, neurosurgery, and psychiatry.* Jul 1992;55(7):546-552.

[12] D'Alessandro G, Di Giovanni M, Roveyaz L, et al. Incidence and prognosis of stroke in the Valle d'Aosta, Italy. First-year results of a community-based study. *Stroke; a journal of cerebral circulation.* Dec 1992;23(12):1712-1715.

[13] Lauria G, Gentile M, Fassetta G, et al. Incidence and prognosis of stroke in the Belluno province, Italy. First-year results of a community-based study. *Stroke; a journal of cerebral circulation.* Oct 1995;26(10):1787-1793.

[14] Di Carlo A, Inzitari D, Galati F, et al. A prospective community-based study of stroke in Southern Italy: the Vibo Valentia incidence of stroke study (VISS). Methodology, incidence and case fatality at 28 days, 3 and 12 months. *Cerebrovascular diseases.* 2003;16(4):410-417.

[15] D'Alessandro G, Bottacchi E, Di Giovanni M, et al. Temporal trends of stroke in Valle d'Aosta, Italy. Incidence and 30-day fatality rates. *Neurological sciences : official journal of the Italian Neurological Society and of the Italian Society of Clinical Neurophysiology.* Feb 2000;21(1):13-18.

[16] Vemmos KN, Bots ML, Tsibouris PK, et al. Stroke incidence and case fatality in southern Greece: the Arcadia stroke registry. *Stroke; a journal of cerebral circulation.* Feb 1999;30(2):363-370.

[17] Kolominsky-Rabas PL, Sarti C, Heuschmann PU, et al. A prospective community-based study of stroke in Germany--the Erlangen Stroke Project (ESPro): incidence and case fatality at 1, 3, and 12 months. *Stroke; a journal of cerebral circulation.* Dec 1998;29(12):2501-2506.

[18] Kubo M, Kiyohara Y, Kato I, et al. Trends in the incidence, mortality, and survival rate of cardiovascular disease in a Japanese community: the Hisayama study. *Stroke; a journal of cerebral circulation.* Oct 2003;34(10):2349-2354.

[19] Morikawa Y, Nakagawa H, Naruse Y, et al. Trends in stroke incidence and acute case fatality in a Japanese rural area : the Oyabe study. *Stroke; a journal of cerebral circulation.* Jul 2000;31(7):1583-1587.

[20] Inagawa T, Ohbayashi N, Takechi A, Shibukawa M, Yahara K. Primary intracerebral hemorrhage in Izumo City, Japan: incidence rates and outcome in relation to the site of hemorrhage. *Neurosurgery.* Dec 2003;53(6):1283-1297; discussion 1297-1288.

[21] Drury I, Whisnant JP, Garraway WM. Primary intracerebral hemorrhage: impact of CT on incidence. *Neurology.* May 1984;34(5):653-657.

[22] Inagawa T. What are the actual incidence and mortality rates of intracerebral hemorrhage? *Neurosurgical review.* Aug 2002;25(4):237-246.

[23] Tanaka H, Ueda Y, Date C, et al. Incidence of stroke in Shibata, Japan: 1976-1978. *Stroke; a journal of cerebral circulation.* Jul-Aug 1981;12(4):460-466.

[24] Zhang LF, Yang J, Hong Z, et al. Proportion of different subtypes of stroke in China. *Stroke; a journal of cerebral circulation.* Sep 2003;34(9):2091-2096.

[25] Labovitz DL, Halim A, Boden-Albala B, Hauser WA, Sacco RL. The incidence of deep and lobar intracerebral hemorrhage in whites, blacks, and Hispanics. *Neurology.* Aug 23 2005;65(4):518-522.

[26] Sacco RL, Boden-Albala B, Gan R, et al. Stroke incidence among white, black, and Hispanic residents of an urban community: the Northern Manhattan Stroke Study. *American journal of epidemiology.* Feb 1 1998;147(3):259-268.

[27] Kissela B, Schneider A, Kleindorfer D, et al. Stroke in a biracial population: the excess burden of stroke among blacks. *Stroke; a journal of cerebral circulation.* Feb 2004;35(2): 426-431.

[28] Minelli C, Fen LF, Minelli DP. Stroke incidence, prognosis, 30-day, and 1-year case fatality rates in Matao, Brazil: a population-based prospective study. *Stroke; a journal of cerebral circulation.* Nov 2007;38(11):2906-2911.

[29] Khan FA, Engstrom G, Jerntorp I, Pessah-Rasmussen H, Janzon L. Seasonal patterns of incidence and case fatality of stroke in Malmo, Sweden: the STROMA study. *Neuroepidemiology.* 2005;24(1-2):26-31.

[30] Thrift AG, Dewey HM, Macdonell RA, McNeil JJ, Donnan GA. Incidence of the major stroke subtypes: initial findings from the North East Melbourne stroke incidence study (NEMESIS). *Stroke; a journal of cerebral circulation.* Aug 2001;32(8):1732-1738.

[31] Thrift AG, Dewey HM, Sturm JW, et al. Incidence of stroke subtypes in the North East Melbourne Stroke Incidence Study (NEMESIS): differences between men and women. *Neuroepidemiology.* 2009;32(1):11-18.

[32] McNaughton H, Feigin V, Kerse N, et al. Ethnicity and functional outcome after stroke. *Stroke; a journal of cerebral circulation.* Apr 2011;42(4):960-964.

[33] Feigin V, Carter K, Hackett M, et al. Ethnic disparities in incidence of stroke subtypes: Auckland Regional Community Stroke Study, 2002-2003. *Lancet neurology.* Feb 2006;5(2):130-139.

[34] Anderson CS, Carter KN, Hackett ML, et al. Trends in stroke incidence in Auckland, New Zealand, during 1981 to 2003. *Stroke; a journal of cerebral circulation.* Oct 2005;36(10):2087-2093.

[35] Aarhus M, Helland CA, Wester K. Differences in anatomical distribution, gender, and sidedness between ruptured and unruptured intracranial aneurysms in a defined patient population. *Acta neurochirurgica.* Dec 2009;151(12):1569-1574.

[36] Unruptured intracranial aneurysms--risk of rupture and risks of surgical intervention. International Study of Unruptured Intracranial Aneurysms Investigators. *The New England journal of medicine.* Dec 10 1998;339(24):1725-1733.

[37] Qureshi AI, Bliwise DL, Bliwise NG, Akbar MS, Uzen G, Frankel MR. Rate of 24-hour blood pressure decline and mortality after spontaneous intracerebral hemorrhage: a

http://dx.doi.org/10.5772/58469

retrospective analysis with a random effects regression model. *Critical care medicine.* Mar 1999;27(3):480-485.

[38] Kawase K, Okazaki S, Toyoda K, et al. Sex difference in the prevalence of deep-vein thrombosis in Japanese patients with acute intracerebral hemorrhage. *Cerebrovascular diseases.* 2009;27(4):313-319.

[39] Wagner I, Volbers B, Kloska S, Doerfler A, Schwab S, Staykov D. Sex differences in perihemorrhagic edema evolution after spontaneous intracerebral hemorrhage. *European journal of neurology : the official journal of the European Federation of Neurological Societies.* Nov 2012;19(11):1477-1481.

[40] Feldmann E, Broderick JP, Kernan WN, et al. Major risk factors for intracerebral hemorrhage in the young are modifiable. *Stroke; a journal of cerebral circulation.* Sep 2005;36(9):1881-1885.

[41] Kimura Y, Takishita S, Muratani H, et al. Demographic study of first-ever stroke and acute myocardial infarction in Okinawa, Japan. *Internal medicine.* Sep 1998;37(9): 736-745.

[42] Zia E, Engstrom G, Svensson PJ, Norrving B, Pessah-Rasmussen H. Three-year survival and stroke recurrence rates in patients with primary intracerebral hemorrhage. *Stroke; a journal of cerebral circulation.* Nov 2009;40(11):3567-3573.

[43] Nakamura T, Xi G, Keep RF, et al. Effects of endogenous and exogenous estrogen on intracerebral hemorrhage-induced brain damage in rats. *Acta neurochirurgica. Supplement.* 2006;96:218-221.

[44] Ye Z, Xie Q, Xi G, Keep RF, Hua Y. Effects of gender on heart injury after intracerebral hemorrhage in rats. *Acta neurochirurgica. Supplement.* 2011;111:119-122.

[45] Nakamura T, Hua Y, Keep RF, Park JW, Xi G, Hoff JT. Estrogen therapy for experimental intracerebral hemorrhage in rats. *Journal of neurosurgery.* Jul 2005;103(1): 97-103.

[46] Auriat A, Plahta WC, McGie SC, Yan R, Colbourne F. 17beta-Estradiol pretreatment reduces bleeding and brain injury after intracerebral hemorrhagic stroke in male rats. *Journal of cerebral blood flow and metabolism : official journal of the International Society of Cerebral Blood Flow and Metabolism.* Feb 2005;25(2):247-256.

[47] Gu Y, Xi G, Liu W, Keep RF, Hua Y. Estrogen reduces iron-mediated brain edema and neuronal death. *Acta neurochirurgica. Supplement.* 2010;106:159-162.

[48] Chen TY, Tsai KL, Lee TY, Chiueh CC, Lee WS, Hsu C. Sex-specific role of thioredoxin in neuroprotection against iron-induced brain injury conferred by estradiol. *Stroke; a journal of cerebral circulation.* Jan 2010;41(1):160-165.

[49] Chen Z, Xi G, Mao Y, Keep RF, Hua Y. Effects of progesterone and testosterone on ICH-induced brain injury in rats. *Acta neurochirurgica. Supplement.* 2011;111:289-293.

[50] Lei B, Mace B, Bellows ST, et al. Interaction Between Sex and Apolipoprotein E Genetic Background in a Murine Model of Intracerebral Hemorrhage. *Translational stroke research.* Jul 1 2012;3(1):94-101.

[51] Bamford J, Sandercock P, Dennis M, et al. A prospective study of acute cerebrovascular disease in the community: the Oxfordshire Community Stroke Project 1981-86. 1. Methodology, demography and incident cases of first-ever stroke. *Journal of neurology, neurosurgery, and psychiatry.* Nov 1988;51(11):1373-1380.

[52] Fogelholm R, Murros K, Rissanen A, Avikainen S. Long term survival after primary intracerebral haemorrhage: a retrospective population based study. *Journal of neurology, neurosurgery, and psychiatry.* Nov 2005;76(11):1534-1538.

[53] Lavados PM, Sacks C, Prina L, et al. Incidence, 30-day case-fatality rate, and prognosis of stroke in Iquique, Chile: a 2-year community-based prospective study (PISCIS project). *Lancet.* Jun 25-Jul 1 2005;365(9478):2206-2215.

[54] Lavados PM, Sacks C, Prina L, et al. Incidence of lobar and non-lobar spontaneous intracerebral haemorrhage in a predominantly Hispanic-Mestizo population--the PISCIS stroke project: a community-based prospective study in Iquique, Chile. Neuroepidemiology. 2010;34(4):214-221.

Chapter 3

The Pathogenesis of Edema and Secondary Insults after ICH

Gaiqing Wang

Additional information is available at the end of the chapter

http://dx.doi.org/10.5772/58542

1. Introduction

Intracerebral hemorrhage (ICH) is a type of acute stroke characterized by extravasation of blood into brain parenchyma and formation of hematoma, leading to edema and tissue damage in the brain. During ICH, rapid accumulation of blood within brain parenchyma leads to disruption of normal anatomy and increased local pressure. Depending on the dynamic of hematoma expansion (growth), the primary damage occurs within minutes to hours from the onset of bleeding and is primarily the result of mechanical damage associated with the mass effect, which compress adjacent tissues, thus destroying them. The 'mass effect' is an important factor in the pathogenic events in ICH. But it may be difficult to predict and manage this effect directly by drug therapies [1-2].

Once present, ICH causes both primary and secondary injury. The primary insult is due to disruption of adjacent tissue and mass effect. Secondary injury occurs with the development of edema, free radical formation, inflammation, and direct cellular toxicity due to the deposited hematoma and subsequent degradation byproducts [3].

2. Body

After arteriolar rupture and parenchymal hemorrhage in the brain, a combination of local compression, cytotoxic injury, inflammation, and surrounding edema ensues. Many patients with ICH deteriorate progressively with no sign of hematoma expansion, suggesting that secondary damage following ICH plays a critical role in neurological deterioration. Secondary damage is, for the most part, attributable to the presence of intraparenchymal blood and may be dependent on the initial hematoma volume, patients age, or ventricular volume. Several lines of evidence show that secondary damage involves blood constituents such as thrombin and hemoglobin as well as its degradation products, which exert biological actions or toxic influences on brain cells. These events, primarily resulting from blood extravasation, which subsequently activates cytotoxic, oxidative and inflammatory pathways, also triggers secondary reactions in the brain parenchyma including recruitment of additional proteases, cerebral edema, and cellular apoptosis, ultimately leading to blood–brain barrier disruption and massive brain cell death. The toxic effects of extravasated blood result mainly from blood components, including red blood cells (RBCs), plasma proteins, coagulation factors, inflammatory mediators , complement components and immunoglobulins. After ICH, the extravasated blood components (primarily erythrocytes and plasma proteins) and the damage-associated molecular patterns, impose a strong cytotoxic, pro-oxidative, and proinflammatory insult toward adjacent viable brain cells and could be seen as early as minutes after onset of ICH [1-3].

In addition to the hematoma, the associated edema may also contribute to the initial neurological deficit, subsequent decline, or death. The edema related to ICH has been cited as a reason for neurological deterioration after the first 24 to 48 h from the onset of symptoms, and it has, to a lesser degree, also been implicated with deterioration as late as 3 weeks. The edema has been demonstrated to be predominately vasogenic with a cytotoxic component. The vasogenic edema is a consequence of blood brain barrier (BBB) breakdown. In the normal brain, the BBB prevents the flow of water into the brain due to hydrostatic pressure gradients. However, when the BBB is disrupted as occurs in ICH, the imbalance in hydrostatic forces result in the entry of an exudative proteinaceous fluid onto the brain parenchyma. The disruption in the BBB is likely a consequence of an inflammatory cascade with resultant expression of specific cytokines and other markers of inflammation. The presence of red blood cells and their subsequent lysis and release of oxyhemoglobin may contribute to the leakage of the BBB. The hemorrhage itself also induces the production of thrombin and the overexpression of matrix metalloproteinases. Thrombin has been demonstrated to be an important factor in the modulation of BBB breakdown. Thrombin may be a major mediator of ICH-induced tumor necrosis factor-α production and an increase of perihematomal tumor necrosis factor-α levels contributes to brain edema formation after ICH. Matrix metalloproteinases also promote BBB disruption and have been associated with increased edema volume via extracellular matrix proteolysis, basal lamina destruction, and the degradation of c-fibronectin [4].

3. Blood plasma components/products

At early stage following ICH, the toxicity of extravasated blood plasma components including blood derived coagulation factors, complement components, immunoglobulins, and other bioactive molecules are proposed to act as contributors to ICH-affected tissue damage.

http://dx.doi.org/10.5772/58542

3.1. Thrombin [1-2]

Thrombin, a serine protease produced rapidly after ICH onset, plays a pivotal role in the blood coagulation cascade. In response to bleeding, a complex series of clotting-factor interactions leads to its conversion by thromboplastin to thrombin, which transforms fibrinogen in plasmainto fibrin, as well as catalyzing many other coagulation-related reactions. As part of its activity in the coagulation cascade, thrombin also promotes platelet activation and aggregation via activation of protease-activated receptors on the cell membrane of the platelet. The primary purpose of thrombin is to stop bleeding as soon as possible and prevent hematoma expansion. Besides its physiological role, substantial lines of evidence indicate that thrombin participates in various pathological conditions in the brain, which contributes to edema formation and blood–brain barrier damage in early brain injury, and activates the cytotoxic, excitotoxic and inflammatory pathways that are involved in secondary injury following ICH.

In the case of ICH, a large amount of blood-derived thrombin invades the brain tissue and exerts biological actions through its proteolytic activity. The substrates for thrombin include proteinase-activated receptors that transduce intracellular signals via trimeric G proteins.The role of thrombin in ICH pathogenesis was first suggested by its possible involvement in edema formation. That is, injection of whole blood into the striatum of rats induced edema, which was prevented by addition of hirudin, a thrombin inhibitor. Edema induced by autologous blood injection into the striatum was attenuated also by argatroban, another inhibitor of thrombin, even when the drug was systemically administered from 6 h after blood injection. ICH-associated edema results from disruption of the blood–brain barrier and death of brain parenchymal cells, both of which may be induced by thrombin.Application of thrombin to rat corticostriatal cultures induced delayed neuron death in the cortical region and shrinkage of the striatal region. Various pharmacological examinations revealed distinct properties of the mechanisms of injury between the cerebral cortex and the striatum. For example, extracellular signal-regulated kinase (ERK), p38, and c-Jun N-terminal kinase (JNK), three major members of mitogen-activated protein kinase (MAPK) family, all contributed to thrombin-induced injury of the striatum, whereas ERK, but not p38 or JNK, was involved in cortical injury. In addition, depletion of microglia from slice cultures rescued striatal tissue, but not cortical cells, from thrombin-induced injury, suggesting that microglia participate only in striatal tissue injury. Involvement of MAPKs and activated microglia in striatal tissue injury was confirmed by an in vivo study where thrombin was directly injected into the striatum of adult rats. On the other hand, plasminogen was found to cooperate with thrombin in inducing cortical injury but not striatal injury.Thrombin-mediated cellular injury may also be mediated by activation of matrix metalloproteinase (MMP)-9. That is, concurrent application of MMP-9 exacerbates cytotoxicity of thrombin in neurons in primary culture, and thrombin cytotoxicity is partially attenuated by MMP inhibitors. Moreover, brain damage induced by autologous blood injection is synergistically attenuated by deletion of the gene encoding MMP-9 and administration of the thrombin inhibitor hirudin.

So, Thrombin can be a double-edged sword, It should be protective during the early stage in ICH to rapidly stop bleeding and prevent hematoma enlargement , whereas its augment in the late stage may result in edema formation and be toxic.

3.2. Haptoglobin (Hp) and Hemopexin(Hx) [1-2]

Haptoglobin and Hemopexin Acts to Combat Hb/Heme Toxicity After ICH.

Haptoglobin (Hp), an acute phase protein, is an abundant blood plasma component that is normally synthesized and released into blood circulation primarily by hepatocytes. The primary function of Hp in blood is to bind and neutralize nephrotoxic free Hb in case of intravascular hemolysis.

Under normal circumstances, Hp represents an effective mechanism by which our body is protected from Hb toxicity. However, because Hp synthesis is not increased by low Hp levels and Hp is not recycled by macrophages, it may take 5 to 7 days for the Hp level to recover if completely sequestered by Hb. Thus, massive hemolysis may lead to persistent hypohaptoglobinemia. Interestingly, Recently study demonstrated that Hp is also produced locally in rat brain after ICH and its expression is significantly increased around the hematoma within hours from the onset of ICH.

The brain-derived Hp appeared to be synthesized and released by oligodendrocytes. Because oligodendroglia are abundant in white matter and are present throughout the gray matter, local production of Hp by these cells likely represents an important endogenous mechanism protecting brain against the extravascular Hb toxicity. Indirect support for such a claim includes (1) primary oligodendrocytes protect neurons in culture from Hb toxicity via Hp release; (2) animals made hypohaptoglobinemic with repetitive Hb administration, before ICH, experience more extensive brain damage; (3) mice genetically engineered to overexpress Hp are less susceptible to ICH injury; and (4) Hp-deficient mice are more vulnerable to ICH injury. In the context of therapeutic relevance, there have been determined that pharmacological intervention with sulforaphane, a naturally occurring agent that acts as NF-E2–related factor-2 (Nrf2) transcription factor activator, increases Hp in blood plasma and brain, and notably reduces brain damage in animal models of ICH. The relevance of Hp genotype as a factor modifying the outcome after ICH has not been studied to date. It could also be relevant for this review to indicate that in addition to the Hp-Hb/CD163 scavenging system, an independent system exists to help remove Hb breakdown products, heme, and iron.

Hemopexin (Hx) is a blood plasma glycoprotein synthesized primarily by hepatocytes. Hx has been shown to bind to heme with a high affinity and forms stable Hx-heme complexes. The heme–Hx complexes are readily endocytosed by macrophages expressing CD91 macroglobulin receptor, (also known as low-density lipoprotein receptor-related protein-1). Although under physiological conditions CD91 plays a role in recycling iron in response to extravascular hemolysis in hematoma-affected tissue, the Hx-heme/CD91 system may facilitate removal of prooxidative heme by icroglia/macrophages.

Recent studies and ongoing research in this laboratory support this notion and suggest that Hx-deficient mice experience augmented ICH injury. More studies of the role of Hx in ICH are warranted.

4. Red blood cells degradation products [1-2, 5-6]

Aside from proteases and plasma products , important blood constituents contributing to ICH pathogenesis are erythrocyte and its degradation products. Following ICH, a large number of red blood cells penetrate into brain parenchyma. Hemolysis does not occur promptly after ICH, but rather, proceeds slowly, taking several days to weeks. Packed RBCs do not cause acute edema development after infusion into the basal ganglia or frontal white matter. However, they do cause delayed edema that appears to be related to release of hemoglobin. Usually, most RBCs start to lyse several days after ICH, but RBC lysis can occur very early.A cascade of events triggered by erythrocyte lysis is critical for the delayed development of edema and the secondary brain damage after ICH.

4.1. Hemoglobin (Hb) and Heme and heme oxygenase-1 (HO-1) [1-2, 5-6]

Red blood cell lysis lead to release of cytotoxic hemoglobin (Hb) with further deterioration of the pathological status quo . Hb and its degradation products, heme and iron, directly compromise the well-being of neighboring brain cells. Hb and heme are potent cytotoxic chemicals capable of causing death to many brain cells.

Hemoglobin(Hb) is a major component of blood and a potent mediator of oxidative stress after ICH. After a cerebral hemorrhage, large numbers of hemoglobin-containing red blood cells are released into the brain's parenchyma and/or subarachnoid space. Hemoglobin is released from lysed red blood cells, heme is liberated from hemoglobin, and hemoglobin as well as heme is taken up by brain parenchymal cells such as microglia and neurons. Prominently, the mechanism of hemoglobin toxicity is via generating free radicals (mainly through Fenton-type mechanism) and massive oxidative damage to proteins, nucleic acids, carbohydrates, and lipids. Heme itself or its degradation products may contribute to formation of brain edema and secondary brain injury after hemorrhagic insults. Hemin, the oxidative form of heme, plays a critical role in Hb-induced brain injury following ICH. Hemin exerts its neurotoxic effects via release of excessive iron, depletion of glutathione and production of free radicals.

Breakdown of the heme moieties of hemoglobin is catalyzed by heme oxygenase-1 (HO-1) into iron, carbon monoxide, and biliverdin, the latter two of which are thought to mediate the anti-inflammatory and antioxidant actions. HO-1, the rate-limiting enzyme for heme catabolism and iron production, after induction of ICH by collagenase or injection of autologous blood, robust expression of HO-1 is induced predominantly in non-neural cells such as microglia/ macrophages and endothelial cells. The role of HO-1 following ICH is controversial.

Under normal conditions, HO-1 is expressed at a very low level, but it is rapidly induced by hemoglobin, heme, and various oxidants. HO-1 has a convincing role in the regulation of intracellular iron. HO-1increased heme catabolism, so it plays a protective role against oxidative injuries in the vascular system.HO-1 induction protects astrocytes from the oxidative toxicity of hemoglobin and heme. HO-1 has a protective role as an intrinsic factor against oxidative stress and the protection depends on the degree of oxidative stress by generating antioxidant bilirubin and vasodilating carbon monoxide. Deficiency of HO-1 in humans and in an HO-1 knockout mouse model leads to vulnerability to oxidant stress and inflammation. In the present study, intravenous administration of nicaraven (a marked synergistic induction of HO-1 protein, for 2 days after subarachnoid hemorrhage) ameliorated delayed cerebral vasospasm in rat subarachnoid hemorrhage models. These results suggest that the enhanced HO-1 expression through a combination of pathological state and pharmacological agent could be an effective strategy to improve the prognosis of heme- and oxidative stress-induced diseases. However, contradictory results have been reported in other researches. Overexpression of HO-1 may be cytotoxic when excessive free iron exceeds the antioxidant properties of heme derived biliverdin. Earlier studies on ICH models in pigs and rabbits demonstrated that tin-mesoporphyrin, an isoform-nonselective HO inhibitor, attenuated edema formation and neuron loss. HO-1 and peroxidized lipid content were significantly higher in CSF from SAH patients with vasospasm, compared with nonvasospasm SAH CSF and correlated with occurrence of vasospasm. It is suggested that bilirubin and HO-1 was induced after hemorrhagic stroke, reflecting the intensity of oxidative stress. One potential explanation for such discrepancy is that HO-1 deficiency could reduce the excessive liberation of free iron from erythrocytes in hematoma (feature that is unique to ICH) and consequently limit the iron-mediated oxidative stress.After some controversy over the beneficial or toxic roles of HO-1, a better understanding of the pivotal HO-1 has evolved. We recently proposed a novel role of HO-1 in ICH pathogenesis. We concluded that upregulation of HO-1 after ICH may be a double-edged sword. The early mild upregulation possibly fit with the events and its overexpression in the late stage may result in its dysfunction and be toxic. It should be prudent to intervene ICH with the inhibitor or activator of HO-1 and think over its potential dual effects.

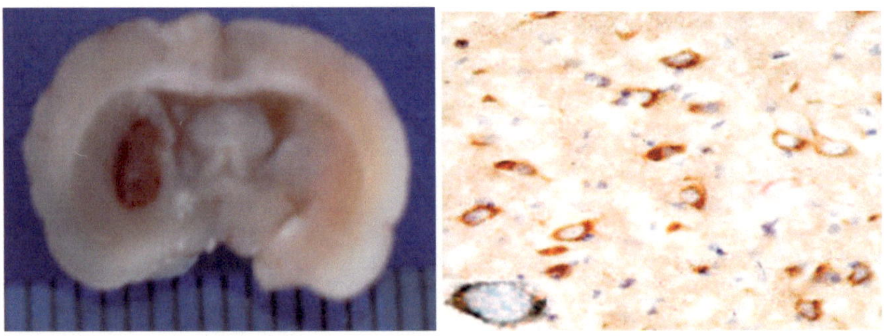

Figure 1. Cerebral slice with autologus blood HO-1 positive staining in ICH rat in ICH rat

4.2. Iron deposits and AQP4 [1-2, 5-6]

As indicated, toxicity of free iron originated from extravascular hemolysis, and HO-mediated catabolism is well-documented. Iron derived from heme degradation may also play a key role in ICH pathogenesis, presumably via acceleration of oxidative stress. After autologous blood injection in rats, induction of HO-1 is followed by a gradual increase in tissue levels of non-heme iron. Iron concentrations in the brain can reach very high levels following RBC lysis, possibly contributing to acute brain edema formation (1st week) and delayed brain atrophy (1 month later). Lysis of RBCs and iron overload contribute to delayed edema formation after ICH. Iron accumulation in brain tissue is toxic and may result in brain damage after ICH. A recent clinical study showed that high serum levels of ferritin, a soluble protein for iron storage, are associated with poor outcome in ICH patients, suggesting that iron is involved in brain damage by ICH. Consistent with this idea, several studies on animal models of ICH showed that the iron chelator deferoxamine attenuated tissue damage and neurological deficits. For instance, intraperitoneal administration of deferoxamine starting from 2 or 6 h after autologous blood injection lessened brain edema, oxidative stress, and motor function deficits in rats. Another study on aged rats showed that deferoxamine was effective in reducing brain edema and motor dysfunction even when administration of the drug was delayed and 48 h, respectively, after autologous blood injection. However, the therapeutic efficacy of deferoxamine might depend on the type of ICH model. Reports mentioned above demonstrating beneficial effects of deferoxamine are based on the model made by autologous blood injection, whereas deferoxamine did not improve the outcome of the collagenase- induced ICH model in rat. On the other hand, a study on clioquinol, another kind of ferrous iron chelator, has demonstrated that oral administration of the drug, starting 6 h after induction of hemorrhage near the internal capsule by collagenase, alleviated motor dysfunction of rats . It recently has been demonstrated that estrogen reduces ferrous iron toxicity in vivo and in vitro, indicating that gender difference in susceptibility to ICH may, in part, be associated with differences in handling ferrous iron toxicity. Iron (II), by reacting with H2O2 generates hydroxyl radicals and clioquinol, by forming stable complexes with ferrous iron prevents its engagement in oxidative reactions. So , all these data suggest that iron particluary ferrous iron (II) plays an important part in brain injury after ICH.

Iron has the potential to mediate a number of deleterious reactions both in vitro and in vivo. Iron accumulation in tissues, particularly if the labile iron pool is increased, is associated with tissue damage.Iron overload in the brain can cause free-radical formation and oxidative damage such as lipid peroxidation after ICH. Brain cells, including neurons, astrocytes, and microglia, show a decreased ability to respond to oxidative stress, particularly with respect to their levels of glutathione and lutathione peroxidase, such that alteration in their iron status may predispose them to iron-induced oxidative stress.

Among the aquaporins(AQP) family, a major water-channel in the central nervous system(CNS)is AQP4, which is a key molecule for maintaining water balance, and its dysfunction or structural damage may cause brain edema.Aquaporin 4, a major water channel protein that is expressed in the brain, plays a key role in the maintenance of brain water homeostasis. It has been proposed that AQP4 may play an important role in the formation of cerebral edema. Because of restricted space within the cranium, salt and water fux in the CNS must be strictly regulated to maintain neuronal functions of the brain. In the CNS, most of the AQP4 is expressed in perimicrovessel astrocyte foot processes, and alterations in AQP4 expression are associated with perturbations of brain water homeostasis. The pattern of AQP4 expression was correlated with blood-brain barrier permeability, which was assessed using contrast enhanced Computed Tomography scanning. Our results showed that AQP4 was mainly located around blood vessels. The current study provided more direct evidence that AQP4 in perivascular astroglial end feet plays a key role in exchange of water between brain, blood, and cerebrospinal fuid. Upregulation of AQP4 induced by iron overload may cause an increased permeability to water in astrocytic membranes. The faint positive immunoreactivity of AQP4 possibly prevent the astrocytes from swelling. So, AQP4 in the brain may be viewed as a final common pathways of cerebral edema.

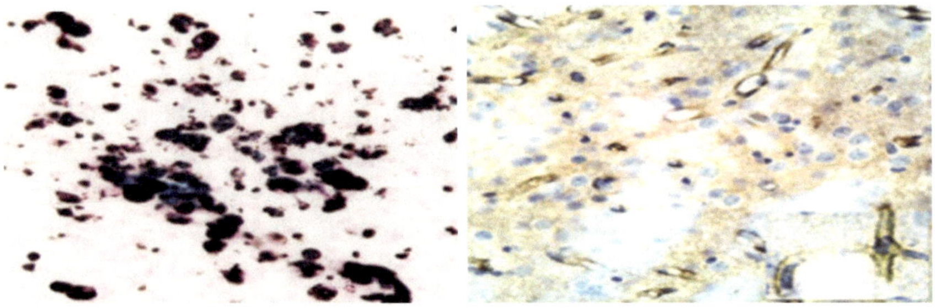

Figure 2. Iron staining after ICH (Perl's staining) AQP4 positive staining in ICH rat

5. Inflammatory mediators [1-4]

Whereas inflammatory mediators generated locally in response to brain injury have the capacity to augment damage caused by ICH (secondary injury), the involvement of inflammatory cells, eg, microglia/macrophages, is vital for removal or cleanup of cellular debris from hematoma, the source of ongoing inflammation. The timely removal of damaged tissue is essential for reducing the length of deleterious pathological process and thereby allowing for faster and more efficient recovery.

Several lines of evidence showed that activation of innate immunity and inflammatory responses contributes to the pathogenesis of secondary injury after ICH. An inflammatory response occurs after ICH, which aggravates ICH-induced brain injury, leading to further tissue damage, blood–brain barrier disruption and edema. The inflammatory mechanisms involved in progression of ICH-induced brain injury include activation of microglial cells, infiltration of inflammatory cells and production of cytokines and chemokines.

5.1. Microglias activation [1-4, 7]

The brain-resident phagocytes, microglia, are highly abundant (10%–15% of total glial cells) in brain and become readily activated within minutes after ICH. Microglia are a type of glial cell that are the resident macrophages of the brain and spinal cord, and thus act as the first and main form of active immune defense in the central nervous system(CNS). Microglial cells fulfill a variety of different tasks within the CNS mainly related to both immune response and maintaining homeostasis.

Microglia constitute 10-15% of the total glial cell population within the brain. Microglia (and astrocytes) are distributed in large non-overlapping regions throughout the brain and spinal

http://dx.doi.org/10.5772/58542

cord. Microglia are constantly scavenging the CNS for plaques, damaged neurons and infectious agents. The brain and spinal cord are considered "immune privileged" organs in that they are separated from the rest of the body by a series of endothelial cells known as the blood–brain barrier, which prevents most infections from reaching the vulnerable nervous tissue. In the case where infectious agents are directly introduced to the brain or cross the blood–brain barrier, microglial cells must react quickly to decrease inflammation and destroy the infectious agents before they damage the sensitive neural tissue. Due to the unavailability of antibodies from the rest of the body (few antibodies are small enough to cross the blood brain barrier), microglia must be able to recognize foreign bodies, swallow them, and act as antigen-presenting cells activating T-cells. Since this process must be done quickly to prevent potentially fatal damage, microglia are extremely sensitive to even small pathological changes in the CNS.

Microglia can be activated by a variety of factors including: glutamate receptor agonists, pro-inflammatory cytokines, cell necrosis factors, lipopolysaccharide, and changes in extracellular potassium (indicative of ruptured cells). Once activated the cells undergo several key morphological changes including the thickening and retraction of branches, uptake of major histocompatibility complex (MHC) class I/II proteins, expression of immunomolecules, secretion of cytotoxic factors, secretion of recruitment molecules, and secretion of pro-inflammatory signaling molecules (resulting in a pro-inflammation signal cascade). Activated non-phagocytic microglia generally appear as "bushy, " "rods, " or small ameboids depending on how far along the ramified to full phagocytic transformation continuum they are. In addition, the microglia also undergo rapid proliferation in order to increase their numbers. From a strictly morphological perspective, the variation in microglial form along the continuum is associated with changing morphological complexity and can be quantitated using the methods of fractal analysis, which have proven sensitive to even subtle, visually undetectable changes associated with different morphologies in different pathological states.

Microglial cells are activated within minutes after the onset of ICH. The activated microglia release proinflammatory cytokines and chemotactic factors, which help to recruit hemotogeneous inflammatory cells to the ICH injury sites. Activated microglial cells undergo morphological and functional changes that include enlargement and thickening of processes, upregulation of proinflammatory proteins, and behavioral changes, including proliferation, migration and phagocytosis. Timely clearance of the extravasated RBCs by activated microglia/macrophages can provide protection from local damage resulting from RBC lysis. The primary neuroprotective role of activated microglia is to clear the hematoma and damaged cell debris through phagocytosis, providing a nurturing environment for tissue recovery. This is characterized first by the transient (18 hours–4 days) infiltration of neutrophils and then a long-term (1 day–months) presence of hematogenous macrophages. However, accumulating evidence has shown that microglial activation contributes to ICH-induced secondary brain injury by releasing a variety of cytokines, chemokines, free radicals, nitric oxide and other potentially toxic chemicals. In addition, several studies have shown that inhibition of microglial activation reduces brain damages in animal models of ICH. Microglial inhibitors, such as minocycline and microglia/macrophage inhibitory factors (tuftsin fragment 1–3), reduce ICH-induced brain injury and improve neurological function in rodents. Clearly, microglial activation mediates ICH-mediated brain injury. Successful removal of injured cells can reduce secondary damage by preventing discharge of injurious proinflammatory cell contents. Resolution of hematoma and inhibition of inflammation are considered potential targets for ICH treatment.

Activated microglia can be stained via the marker ionized calcium-binding adapter molecule 1 (IBA1), which is upregulated during activation. Microglia are the only cells in the brain to express Iba1.

One way to control neuroinflammation is to inhibit microglial activation. Studies on microglia have shown that they are activated by diverse stimuli but they are dependent on activation of mitogen-activated protein kinase (MAPK). Previous approaches to down-regulate activated microglia focused on immunosuppressants. Recently, minocycline (a tetracycline derivative) has shown down-regulation of microglial MAPK. Another promising treatment is CPI-1189, which induces cell death in a tumor necrosis factor (TNF) α-inhibiting compound that also down-regulates MAPK. Recent study shows that nicergoline (Sermion) suppresses the production of proinflammatory cytokines and superoxide anion by activated microglia. Microglial activation can be inhibited by MIF (microglia/macrophage inhibitory factor, tuftsin fragment 1–3, Thr-Lys-Pro). MIF-treated mice showed reduced brain injury and improved neurologic function in a mouse model of collagenase-induced intracerebral hemorrhage.

Albeit some inflammatory responses generated by microglia/macrophages after ICH may aggravate brain injury, microglia/macrophages-mediated phagocytosis is instrumental in conducting brain clean-up, the process that must occur to allow for tissue repair and functional recovery. A fast and efficient removal of apoptotic, dislocated (eg, extravascular erythrocytes), and damaged cells before the discharge of injurious and proinflammatory cell contents (damageassociated molecular patterns) occurs and may help to reduce secondary damage.

So, by inhibiting the activation of microglial cells, namely the inhibition of brain primary response to the timely removal of damaged tissue and self repairing systems, to reduce the amount of potential damages, the desirability of this way is still questionable .

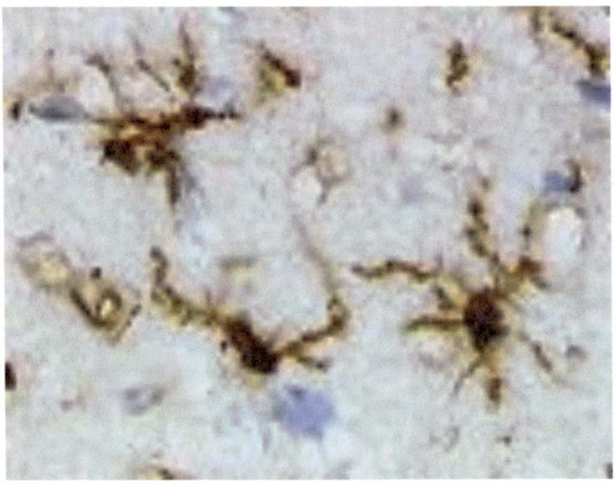

Figure 3. Microglia - ramified form from rat cortex before traumatic brain injury (lectin staining with HRP)

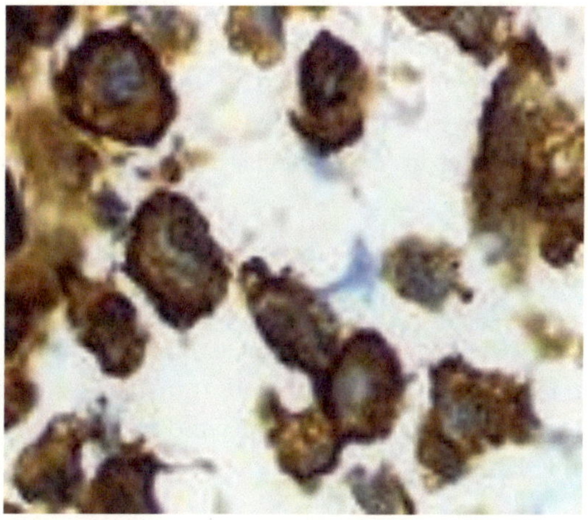

Figure 4. Microglia/Macrophage - activated form from rat cortex after traumatic brain injury (lectin staining with HRP)

5.2. Infiltration of inflammatory cells [1-3]

Besides microglia, other blood-derived inflammatory cells, such as leukocytes and macrophages, are also activated after ICH and contribute to ICH-induced brain injury Neutrophil infiltration occurs less than 1day after the onset of ICH, and the infiltrating neutrophils die by apoptosis within 2days . Neutrophils are believed to contribute to brain injury after ICH. Depletion of neutrophils reduced blood–brain barrier disruption, axon injury and inflammation in a rat model of ICH and was found to prevent tissue plasminogen activator (tPA)-induced ICH in a rat model of cerebral ischemia. Neutrophils may damage brain tissues by producing reactive oxygen species (ROS) and releasing proinflammatory cytokines and matrix metalloproteinases (MMPs). Dying leukocytes can cause further brain injury by stimulating microglia/macrophages to release proinflammatory factors. Activated macrophages are indistinguishable from resident microglia in morphology and function. Similar to activated microglia, activated leukocytes and macrophages release a variety of cytokines, chemokines, free radicals and other potentially toxic chemicals.

5.3. Production of cytokines

Cytokines are well-known to be associated with inflammation and immune activation. Although cytokines are released by many cells, including microglia/macrophages, astrocytes and neurons, the major sources of cytokines are activated microglia/macrophages.

5.3.1. TNF-α and IL-1β [1-3]

Many studies have shown that two major proinflammatory cytokines, TNF-α and interleukin1β (IL-1β), exacerbate ICH-induced brain injury. After ICH, TNF-α is significantly increased both *in vivo* and *in vitro*, which may contribute to brain edema formation and brain injury in animal models of ICH. Consistent with animal studies, clinical studies support the proposition that TNF-α contributes to ICH-induced brain injury. Plasma TNF-α has been shown to correlate with the magnitude of the perihematomal brain edema in patients with ICH. Single-nucleotide polymorphisms in the TNF-α gene promoter are associated with spontaneous deep ICH. Similarly, IL-1β has been found to be upregulated after ICH in an animal model and to produce detrimental effects, including brain edema and blood–brain barrier disruption.

5.3.2. Nuclear factor kappa-B (NF-κB) [1-3]

The apoptotic pathway in ICH may involve nuclear factor-kappa B (NF-κB), which is a ubiquitous transcription factor that, when activated, translocates to the nucleus and binds to DNA. NF-κB is associated with apoptotic cell death and has been reported in the role of cell death after experimental ICH in rats .

The inflammatory signaling involves a coordinated effort of different molecules and cell types and is largely coordinated by a ubiquitous transcription factor, NF-κB, a transcription factor involved in inflammatory responses, is a key regulator of many proinflammatory cytokines, such as TNF-α , IL-1β and MMP-9 are involved in various pathological conditions, including ICH-mediated brain injury. Activation of NF-κB occurs within minutes and lasts for at least 1 week after the onset of ICH. The activity of NF-κB correlates with perilesional cell death after ICH in rats and is positively associated with the progression of apoptotic cell death in patients with ICH. Several lines of evidence have shown that NF-κB is activated by RBCs and plasma via signaling pathways involving free radicals, cytokines and glutamate receptors. Cellular necrosis likely occurs at the core of the hemorrhage; however, apoptosis has been observed in the perihematomal region.

5.3.3. CD36 [1, 3]

Microglia and macrophages express various cell surface receptors, including scavenger receptors (eg, CD36) that assist in phagocytosis / endocytosis - mediated removal of cellular debris after tissue injury, including brain injury after ICH. One specific study evaluated CD36, a class II scavenger receptor that is transcriptionally regulated by peroxisome proliferator-activated receptors (PPARs). This study used in vitro and in vivo models and demonstrated that: (1) microglia/macrophages utilize CD36 to promote phagocytosis of red blood cell; and (2) treating animals with PPAR agonists (eg, rosiglitazone, pioglitazone, or 15D-PGJ2), which increased CD36 expression, results in faster hematoma resolution and improved functional recovery after ICH.

5.3.4. Toll-like receptors (TLRs) [3]

Toll-like receptors (TLRs) is expressed in microglia, the resident macrophages of the brain, belong to a large family of pattern recognition receptors that play a key role in innate immunity and inflammatory responses. It has been reported that TLR4 is upregulated in a rat model of ICH and that its signaling pathway contributes to poor outcome after ICH. TLR4 is activated by many endogenous ligands, such as heme and fibrinogen, which are produced in the brain after ICH. Our recent *in vivo* study shows that activation of TLR4 by heme causes ICH-induced inflammatory injury via the MyD88/TRIF signaling pathway and that effective blockade of TLR4 by its antibody suppresses ICH-induced inflammation. Thus, the TLR4 signaling pathway could be a promising therapeutic target for ICH treatment.

5.3.5. Matrix metalloproteinases(MMPs) [1-2, 4]

The hemorrhage itself also induces the overexpression of matrix metalloproteinases(MMPs). Matrix metalloproteinases also promote BBB disruption and have been associated with increased edema volume via extracellular matrix proteolysis, basal lamina destruction, and the degradation of c-fibronectin.

Matrix metalloproteinases (MMPs) are a large family of calcium-dependent zinc-containing endopeptidases, which are responsible for the tissue remodeling and degradation of the extracellular matrix (ECM), including collagens, elastins, gelatin, matrix glycoproteins, and proteoglycan. Matrix metalloproteinases are excreted by a variety of connective tissue and pro-inflammatory cells including fibroblasts, osteoblasts, endothelial cells, macrophages, neutrophils, and lymphocytes. These enzymes are expressed as zymogens, which are subsequently processed by other proteolytic enzymes (such as serine proteases, furin, plasmin, and others) to generate the active forms. After hemorrhagic events in the brain, several members of the MMP family are recruited and involved in pathogenic processes. An early study on collagenase injection model reported that the MMP inhibitor BB-1101 could reduce brain edema when administered 6h after induction of ICH. In human ICH patients, expression of MMP-9 and MMP-3 increases after the incident. MMP-9 expression is induced in astrocytes and neurons in the perihematomal area, possibly by an action of hemoglobin. Increased MMP-9 is associated with the extent of perihematomal edema, whereas increased MMP-3 is associated with high mortality. Experiments using MMP-9–deficient mice demonstrated that MMP-9 derived from both blood and brain parenchyma contributes to edema formation after autologous blood injection. In addition, MMP-9–deficient mice displayed lower levels of neurodegeneration, neutrophil infiltration, and microglia/macrophage reactions than wild-type mice. In the collagenase injection model, MMP-9 expression was found mainly in neurons and vascular endothelial cells, and administration of the MMP inhibitor GM6001, beginning 2 h after induction of ICH, attenuated neutrophil infiltration, oxidative stress, brain edema, neurodegeneration, and neurological impairment. With regard to MMP-3, early induction of this enzyme may contribute to brain damage in combination with other proteases such as MMP-9 and thrombin. Other lines of evidence indicate that MMP-12 may play a key role in the pathogenesis of ICH. That is, MMP-12 was induced most prominently among MMP isozymes in the collagenase-injection model in rats and mice, and MMP-12 deficient mice showed better functional recovery after ICH as well as reduced levels of recruitment of microglia/macrophages in the perihematomal region.It should be noted that several studies reported conflicting results. For example, a study on MMP-9–deficient mice showed that collagenase injection into the striatum of the mutant mice resulted in enhanced bleeding, increased mortality, and exacerbated neurological deficits. These changes may be attributable to heightened expression of MMP-2 and MMP-3 in response to ICH, together with lowered levels of collagen in the brain of MMP-9–deficient mice . In addition, systemic administration of BB-94, a broad spectrum MMP inhibitor, from 30 min before collagenase injection increased hemorrhagic volume and the number of cells exhibiting DNA fragmentation.

6. Oxidative injury [1-2]

Neurological deficits associated with ICH were also released from RBC lysis is a potent cytotoxic chemical that generates free radicals and oxidative damage, causing death of surrounding cells. ROS are produced after ICH and contribute to ICH pathogenesis. In addition, phagocytosis generates a large amount of ROS that can damage macrophages and neurons. Oxidative stress appears to play a prominent role in ICH pathogenesis. In addition to increased free radical generation, damage to brain tissue may result from the impairment of the endogenous antioxidative enzyme system in response to ICH. Direct evidence for the causal relationship between free radicals and ICH injury was by demonstrating the efficacy of antioxidants as therapeutic agents. Specifically, the free radical scavengers, such as dimethylthiourea, α-phenyl-N-tert-butyl nitrone, NXY-059 (a sulfonyl derivative of α-phenyl-N-tert-butyl nitrone) or deferoxamine, a drug chelating pro-oxidative iron, significantly reduced brain injury in animal models of ICH. Mice with generically deleted NADPH oxidase, a key enzyme involved in generating ROS, showed reduced damage after ICH.

It reported a significant reduction in the levels of manganese superoxide dismutase and copper/zinc superoxide dismutase, the key enzymes of the antioxidant defense system in brains, after intracerebral injection of lysed erythrocytes.

Nrf2, a key player in antioxidative homeostasis. By binding to the antioxidant response element, Nrf2 regulates the expression of many detoxification and antioxidant enzymes, including superoxide dismutase, catalase, glutathione- Stransferase, glutathione peroxidase, HO-1, NAD(P)H quinine oxidoreductase -1, peroxiredoxin, or thioredoxin. To activate Nrf2 in animals after ICH, researchers utilized a naturally occurring organosulfur compound, sulforaphane. As expected, treatment with sulforaphane effectively increased the expression of Nrf2-regulated antioxidant genes, including catalase, superoxide dismutase, and glutathione-Stransferase, in brain tissue after ICH. Notably, this expression of antioxidants corresponded to reduced oxidative damage to proteins and lipids within the ICH-affected brain and importantly, with less severe neurological deficits. Nrf2 plays critical safeguard function in defending brain against oxidative stress associated with ICH pathogenesis.

7. Glutamate [2]

Glutamate has long been recognized as the major excitatory neurotransmitter in the central nervous system (CNS). This amino acid is also well known as an important player in various

http://dx.doi.org/10.5772/58542

CNS disorders, since over-activation of ionotropic glutamate receptors causes neuronal damage via processes called excitotoxicity. Several lines of evidence suggest that glutamate is involved in the pathogensis of ICH. Transient elevation of the extracellular concentration of glutamate in the perihematomal region was demonstrated in rabbits following injection of autologous blood into the gray matter of the cerebrum. Subsequently, the effect of memantine, a low-affinity blocker of the N-methyl-D-aspartate subtype of glutamate receptor–associated channels, was investigated in the collagenase-injection model in rats. Daily intraperitoneal administration of memantine, starting from 30 min after induction of ICH, reduced hemorrhage volume, apoptotic cell death, neutrophil infiltration, and the number of microglia/macrophages in the periphery of hematoma.

8. Ischemia [8, 4]

Based on the recent publications, several potential factors of secondary ischemic injury after ICH have been consistently associated with acute ischemic lesions around hematoma. Although there is restricted diffusion within the hematoma during the first 2 weeks, an effect of increased viscosity and suspectibility effects from blood breakdown products, much more attention has been paid to the potential for ischemia in the surrounding tissue.

9. Conclusion

In conclusion, blood constituents, including hemoglobin-derived products as well as proteases such as thrombin and Haptoglobin, play important roles in the pathogenic events. Inflammatory reactions involving activated microglia, neutrophils, and production of proinflammatory cytokines also constitute a critical aspect of pathology leading to neurodegeneration and tissue damage. The mechanisms of secondary cerebral injury after ICH are complex and multidisciplinary. From a protective response into the later damaged process, they are interacting and overlapping, so we have to weigh the benefits (positive effects) and risks (adverse effects) carefully when we intervene in ICH.

Author details

Gaiqing Wang*

Address all correspondence to: wanggq08@126.com

Department of Neurology, The Second Hospital, Shanxi Medical University, Shan Xi, China

References

[1] Aronowski J, Zhao X. Molecular pathophysiology of cerebral hemorrhage: secondary brain injury. Stroke. 2011 ;42(6):1781-6.

[2] Katsuki H. Exploring neuroprotective drug therapies for intracerebral hemorrhage. J Pharmacol Sci. 2010;114(4):366-78

[3] Fang H, Wang PF, Zhou Y, Wang YC, Yang QW. Toll-like receptor 4 signaling in intracerebral hemorrhage-inducedinflammation and injury. J Neuroinflammation. 2013; 17(10):27

[4] Sangha N, Gonzales NR. Treatment Targets in Intracerebral Hemorrhage. Neurotherapeutics. 2011;8(3):374-87

[5] Wang G, Yang Q, Li G, Wang L, Hu W, Tang Q, Li D, Sun Z. Time course of heme oxygenase-1 and oxidative stress after experimental intracerebral hemorrhage. Acta Neurochir (Wien). 2011;153(2):319-25.

[6] Qing WG, Dong YQ, Ping TQ, Lai LG, Fang LD, Min HW, Xia L, Heng PY. Brain edema after intracerebral hemorrhage in rats: the role of iron overload and aquaporin4. J Neurosurg. 2009;110(3):462-8.

[7] Wikipedia, the free encyclopedia.microglia. http://en.wikipedia.org/wiki/Microglia/ (accessed 11 Nov 2013).

[8] Prabhakaran S, Naidech AM. Ischemic brain injury after intracerebral hemorrhage: a critical review. Stroke. 2012;43(8):2258-63.

Chapter 4

Cerebral Amyloid Angiopathy

Vikas Chaudhary, Shahina Bano,
Sachchidanand Yadav, Umesh Chandra Garga and
Sachin Kumar Singh

Additional information is available at the end of the chapter

http://dx.doi.org/10.5772/58461

1. Introduction

Cerebral amyloid angiopathy (CAA) is an important but under-recognized cause of spontaneous intracranial hemorrhage (ICH) in the normotensive individuals. [1] Both sporadic and hereditary forms may occur. Hereditary form of CAA is seen at a younger age, as early as the third decade; in contrast, the sporadic form is more common in elderly and increases in both prevalence and severity with increasing age. CAA results from deposition of β-amyloid protein in small and medium sized cortical, subcortical, and leptomeningeal vessels. This deposition is responsible for the wide spectrum of clinical symptoms and neuroimaging findings. [1,2] CAA is not associated with the presence of systemic amyloidosis. [3] Majority of cases of CAA are asymptomatic. However, symptomatic patient may present with sudden neurological deficit due to transient ischemic attack, progressive cognitive decline, or potentially devastating intracranial hemorrhage. [3,4] Computed tomography is the imaging modality of choice for evaluation of suspected acute intracranial hemorrhage. Magnetic resonance imaging is a sensitive technique for identifying microhemorrhages, microangiopathy-related ischemic changes and assessment of disease progression. The early and accurate diagnosis of CAA is important because of the likely implication it has on future management targeted to reduce the risk of recurrent hemorrhage. In this chapter our emphasis will be on the complex pathophysiology, important clinical and radiological features and the role of imaging in secondary prevention of CAA related ICH. [1,3,4]

2. Pathophysiology

CAA is characterized by deposition of β-amyloid protein in media and adventitia of small and medium-sized cortical, subcortical, and leptomeningeal vessels, with sparing of similar-sized vessels in the deep white matter. [2] The complex structural changes in the vessel wall related to β-amyloid deposition include endothelial dysfunction, loss of smooth muscle cells, fibrinoid necrosis, vessel wall fragmentation (fragile vessel) and microaneurysm formation. All these factors predispose the patient to repeated episodes of blood vessel leakage and frank hemorrhages in response to sudden increase in blood pressure or minor trauma. [3,4] CAA related inflammation (termed as cerebral amyloid angitis or cerebral amyloid inflammatory vasculopathy) is typically perivascular and may be associated with frank vasculitis. β-amyloid deposition causes vessel wall thickening and subsequent luminal narrowing leading to ischemic changes. [2] The deposition may also impair the perivascular drainage, leading to dilatation of perivascular spaces (also known as Virchow Robin spaces) within the lobar region and in deep cerebral white matter. The enlarged perivascular spaces, a potential useful neuroimaging marker of CAA, can reach several millimeters in diameter and may be visible on brain imaging. [5,6] Histologically, β-amyloid deposits stained with Congo red show classic yellow-green birefringence under polarized light. [3]

3. Clinical spectrum

In general, hereditary form of CAA has an earlier onset and more severe clinical manifestations than sporadic CAA. Symptomatic CAA has variable clinical presentations, which include sudden neurologic deficit (stroke) related to acute ICH, TIA-like symptoms, cognitive impairment and dementia. [7]

The most common presentation of CAA is the development of a sudden neurological deficit secondary to an acute ICH. Specific clinical symptoms and signs depend on both the size and location of the ICH. CAA can have similar presentation as acute ICH related to other causes: headache, nausea and vomiting, loss of consciousness, focal neurological deficits and seizures. [8]

Transient-ischemic attack (TIA) like symptoms also termed as "amyloid spells" is the next most commonly described presentation. The spells are typically brief (<30mts) and are characterized by recurrent, stereotyped episodes of 'positive' spreading sensory symptoms (paraesthesias). The spells are related to hemorrhagic components of CAA, for example cortical microbleeds (CMBs), cortical subarachnoid hemorrhage (cSAH), or cortical superficial siderosis. [4,9]

The prevalence of CAA is significantly higher in demented patients (due to Alzheimer disease) compared to non-demented patients. CAA-related dementia is slowly progressive, similar to that seen in patients with Alzheimer disease. [10] CAA is also the direct cause of cognitive impairment that progresses rapidly over the course of a few weeks. These patients may present with confusion and disorientation. [3]

http://dx.doi.org/10.5772/58461

4. Neuroimaging correlates of CAA

The Boston criteria (table 1) was first proposed in 1990 in order to standardize the diagnosis of cerebral amyloid angiopathy. They comprise of combined clinical, imaging and pathological parameters. [11]

Definite CAA
Full postmortem examination demonstrating:
-Lobar, cortical, or corticosubcortical hemorrhage
-Severe CAA with vasculopathy
-Absence of other diagnostic lesion
Probable CAA with supporting pathology
Clinical data and pathologic tissue demonstrating:
-Lobar, cortical, or corticosubcortical hemorrhage
-Some degree of CAA in specimen
-Absence of other diagnostic lesion
Probable CAA
Clinical data and MRI or CT demonstrating:
-Multiple hemorrhages restricted to lobar, cortical, or corticosubcortical regions
-Age more than 55 years
-Absence of other cause of hemorrhage
Possible CAA
Clinical data and MRI or CT demonstrating:
-Single lobar, cortical, or corticosubcortical hemorrhage
-Age more than 55 years
-Absence of other cause of hemorrhage
CAA: Cerebral amyloid angiopathy

Table 1. Boston criteria for diagnosis of CAA-related hemorrhage

Recognition of the imaging findings of CAA is important for correct diagnosis. The important imaging correlates of CAA include: (i) large intracranial hemorrhages (ICHs), (ii) cerebral microbleeds (CMBs), (iii) convexity subarachnoid hemorrhages (cSAH), (iv) cortical superficial siderosis, (v) white matter changes (leukoaraiosis), and (vi) prominent VRSs.

The majority of ICHs (>75%) in elderly are spontaneous due to rupture of small arteries affected by either of the two processes; the hypertensive arteriopathy or CAA. Distribution of ICH reflects the underlying microangiopathy. Hypertensive arteriopathy is characterized by lipohyalinosis and fibrinoid necrosis of lenticulostriate perforators located in deep gray nuclei (i. e. basal ganglia, thalami) and infratentorial location (i. e. pons). In contrast, CAA related ICH is preferentially lobar (any lobe may be involved); less commonly involves the cerebellum and rarely the deep nuclei or the brainstem. [12] CAA-related ICH represents 2% of all ICH and is an important cause of hemorrhage in normotensive elderly patients without trauma.

Nonenhanced CT helps to exclude the presence or absence of an acute ICH and provides information regarding the location, size, shape and extension of ICH (Figure 1). MR imaging is most sensitive for detection of chronic hemorrhages in suspected cases of CAA. After ICH, hemosiderin remains stored within the macrophages, leads to focal dephasing of the MR signal, and causes hemosiderin-containing areas to appear dark on T2-weighted spin-echo sequences. This effect may be further enhanced by the use of imaging techniques with high sensitivity for differences in magnetic susceptibility, such as the T2*-weighted gradient-echo (GRE) sequence (Figure 2). The susceptibility-weighted imaging (SWI) has been found to be more sensitive than conventional GRE techniques for the detection of blood products. [13,14,15]

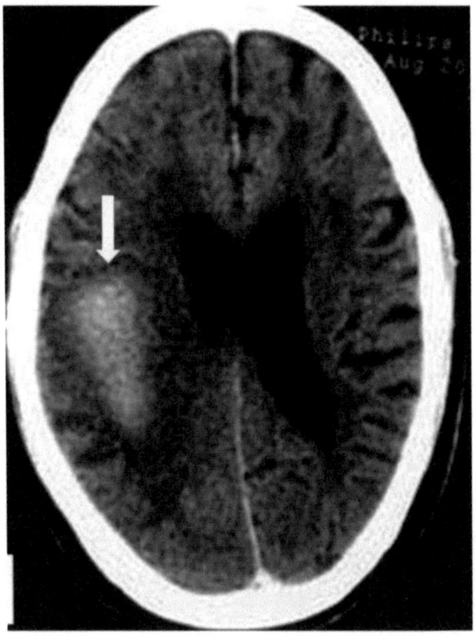

Figure 1. Plain axial CT image of a 66-year-old normotensive male reveals a large, right parietal lobe, hyperdense, sub-acute hemorrhage (arrow) causing mass effect in form of effacement of adjacent sulci and compression of posterior part of the body of ipsilateral lateral ventricle. Diffuse periventricular white matter hypodensities (leukoaraiosis) is also noted.

The hemorrhage is typically lobar and cortical-subcortical in distribution; it generally spares the deep white matter, basal ganglia and the brainstem. Symptomatic ICH is large (>5mm), while microhemorrhages (<5mm) are often clinically silent. CAA-related macrohemorrhages typically exhibit irregular borders (Figure 1) and may be associated with subarachnoid hemorrhage, subdural hemorrhage, or, less commonly, intraventricular hemorrhage. CMBs are small, well demarcated; rounded lesions not detected on conventional MRI (Figure 2). The presence of chronic cortical-subcortical microhemorrhages in association with large ICH increases the probability of CAA (Figure 3). Multiplicity and recurrence of ICH further favors CAA. [16,17]

http://dx.doi.org/10.5772/58461

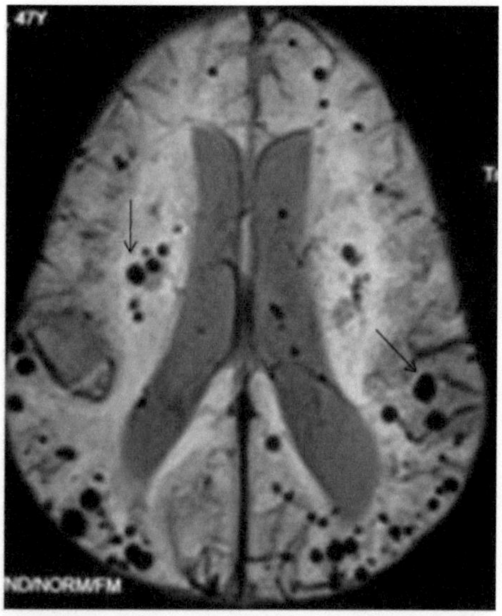

Figure 2. T2*-weighted gradient-echo MR image of a 62-year-old male patient shows diffuse and small multifocal hemosiderin deposits (chronic microhemorrhages) in corticosubcortical location of both the cerebral hemispheres, seen as signal void areas (blooming) (arrows).

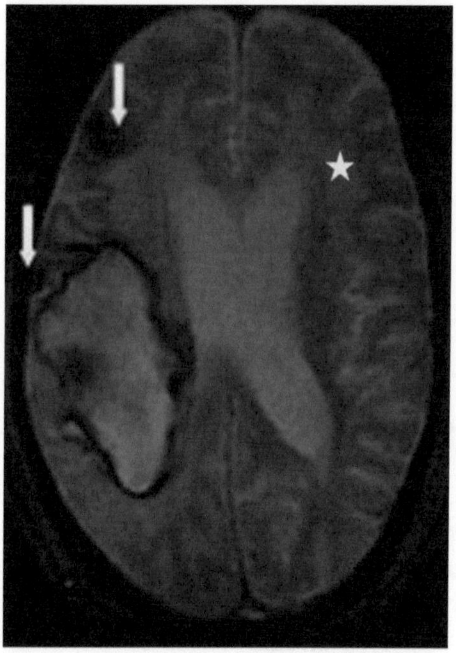

Figure 3. T2*-weighted gradient-echo MR image of same patient as in figure 1, demonstrates a large right parietal hematoma in association with multiple chronic cortical-subcortical microhemorrhages (arrows) and periventricular white matter abnormality (asterisk), thus increasing the probability of CAA.

Chronic subarachnoid hemorrhage (cSAH) and superficial siderosis are quite characteristic of CAA. cSAH often results from lobar ICH extending to the cortical surface (Figure 4). Cortical superficial siderosis describes hemosiderin deposition in the superficial layers of the cerebral cortex and may follow repeated episodes of bleeding in the subarachnoid space. On T2*-weighted GRE sequence, cortical superficial siderosis shows a characteristic gyriform pattern of hypointense signal. [18]

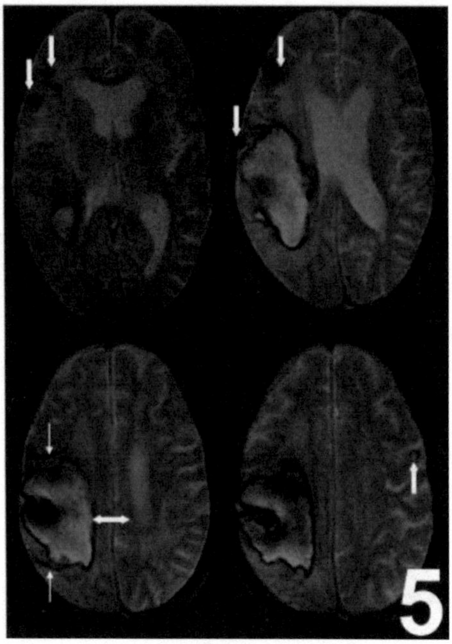

Figure 4. T2*-weighted gradient-echo MR image of same patient as in figure 1, shows two linear areas of signal void (thin arrows) in the vicinity of large primary lobar hemorrhage (thick arrow) suggest chronic subarachnoid hemorrhage and cortical superficial hemosiderosis.

MRI is also sensitive for identifying CAA-related inflammation and ischemic changes and assessment of disease progression. [19] Leukoaraiosis, a radiological term which describes imaging changes in deep cerebral white matter, is a nonspecific finding seen in patients with CAA and can be due to demyelination, infarction or edema. On imaging, leukoaraiosis appears as patchy or confluent, CT hypodense or T2/FLAIR (fluid attenuated inversion recovery) hyperintense white matter abnormality with or without sparing of subcortical U fibers (Figure 5). Leukoaraiosis with sparing of U fibers is secondary to ischemic white matter damage and is seen in association with long standing dementia; whereas, white matter changes that extend to involve the U fibers are common in patients with subacute cognitive decline and are associated with mass effect due to inflammatory edema. [20,21] White matter changes in CAA increases over time and is an important contributor to overall disease burden. The CAA should be considered in the broad differential diagnosis of leukoencephalopathy especially if associated with progressive dementia or cognitive impairment. [13]

http://dx.doi.org/10.5772/58461

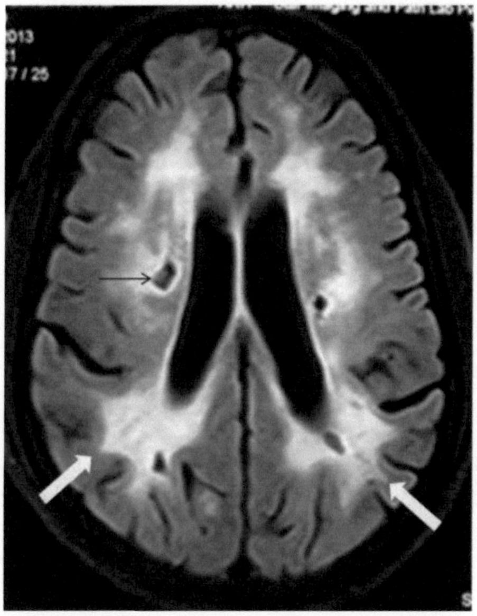

Figure 5. Axial MRI (FLAIR sequence) of the same patient as in figure 2, demonstrates bilateral symmetrical, confluent, supratentorial periventricular white matter hyperintensity with involvement of posterior subcortical U fibers (thick arrows). Also note dilated perivascular spaces (Virchow Robin spaces) in deep cerebral white matter (thin arrow).

5. Management prospect and prognosis

Currently, there is no treatment to halt or reverse β-amyloid deposition. Thus, attention is directed instead to the prevention of adverse outcomes associated with CAA, such as recurrent hemorrhages or progressive dementia. In this context, MRI may help in selecting patients for different types of secondary prevention of stroke. MR evidence of higher number of chronic microbleeds on baseline GRE MR images are predictive of a greater risk of recurrent hemorrhage and future cognitive impairment. A routine use of GRE MRI sequence is suggested to detect microbleeds in older people to avoid potentially dangerous anticoagulant or antiplatelet therapy. [22]

The role of neurosurgery in ICH remains to be defined clearly. However, hematoma evacuation appears relatively safe in patients <75 years of age without intraventricular extension. [23] For future treatment of CAA, it is important to identify patients early in the course of disease before ICH or dementia occurs, to allow the use of disease modifying therapies. [24] Tramiprosate has been found to be a safe treatment option for patients with suspected CAA. Tramiprosate is an ionic compound that binds with soluble β-amyloid, interferes with the amyloid cascade and delays or inhibits the progression of CAA. [25]

6. Summary

- Cerebral amyloid angiopathy (CAA) is an important cause of spontaneous cortical-subcortical intracranial hemorrhage in the normotensive elderly.
- Leukoencephalopathy in conjunction with acute or chronic ICH in a cortical-subcortical location increases the diagnostic specificity for CAA.
- Computed tomography is the imaging study of choice for evaluation of suspected acute cortical hemorrhage.
- MRI is best suited for identification of chronic cortical-subcortical hemorrhages, ischemic sequelae of the disease and assessment of disease progression.
- The burden of asymptomatic cerebral microhemorrhages detectable by GRE MRI in patients with CAA is a good predictor of hemorrhage recurrence, and therefore highlights the importance of secondary prevention in CAA-related PICH.

Author details

Vikas Chaudhary[1*], Shahina Bano[2], Sachchidanand Yadav[2], Umesh Chandra Garga[2] and Sachin Kumar Singh[3]

*Address all correspondence to: dr_vikaschaudhary@yahoo.com

1 Department of Radiodiagnosis, Employees' State Insurance Corporation (ESIC) Model Hospital, Gurgaon, Haryana, India

2 Department of Radiodiagnosis, PGIMER, Dr. Ram Manohar Lohia (RML) Hospital, India

3 Department of Radiodiagnosis, Jawahar Lal Nehru (JLN) Medical College, Bhagalpur, Bihar, India

References

[1] Bano S, Yadav SN, Garga UC, Chaudhary V. Sporadic cerebral amyloid angiopathy: An important cause of cerebral hemorrhage in the elderly. Journal of Neurosciences in Rural Practice 2011;1:87-91.

[2] Vinters HV. Cerebral amyloid angiopathy: a critical review. Stroke 1987;18:311-24.

[3] Chao CP, Kotsenas AL, Broderick DF. Cerebral amyloid angiopathy: CT and MR imaging findings. Radiographics 2006; 26:1517-31.

[4] Andreas C, Qiang G, David JW. Sporadic cerebral amyloid angiopathy revisited: recent insights into pathophysiology and clinical spectrum. J Neurol Neurosurg Psychiatry 2012;83:124-37.

[5] Roher AE, Kuo YM, Esh C, Knebel C, Weiss N, Kalback W, et al. Cortical and leptomeningeal cerebrovascular amyloid and white matter pathology in Alzheimer's disease. Mol Med 2003;9:112-22.

[6] Wardlaw JM. Blood brain barrier and cerebral small vessel disease. J Neurol Sci 2010;299:66-71.

[7] Revesz T, Ghiso J, Lashley T, Plant G, Rostagno A, Frangione B, et al. Cerebral amyloid angiopathies: a pathologic, biochemical, and genetic view. J Neuropathol Exp Neurol 2003;62:885-98.

[8] Rosand J, Greenberg SM. Cerebral amyloid angiopathy. Neurologist 2000;6:315-25.

[9] Greenberg SM, Vonsattel JP, Stakes JW, Gruber M, Finklestein SP. The clinical spectrum of cerebral amyloid angiopathy: presentations without lobar hemorrhage. Neurology 1993;43:2073-9.

[10] Thal DR, Griffin WS, de Vos RA, Ghebremedhin E. Cerebral amyloid angiopathy and its relationship to Alzheimer's disease. Acta Neuropathol 2008;115:599-609.

[11] Kundsen K, Rosand J, Karluk D, Greenberg SM. Clinical diagnosis of cerebral amyloid angiopathy: validation of the Boston criteria. Neurology 2001;56:537-9.

[12] Zhan RY, Tong Y, Shen JF, Lang E, Preul C, Hempelmann RG, et al. Study of clinical features of amyloid angiopathy hemorrhage and hypertensive intracerebral hemorrhage. J Zhejiang Univ Sci 2004;5:1262-9.

[13] Chao PC, Kotsenas LA, Broderick FD. Cerebral amyloid angiopathy: CT and MRI imaging findings. Radiographics 2006;26:1517-31.

[14] Blitstein MK, Tung GA. MRI of cerebral microhemorrhages. AJR Am J Roentgenol 2007;189:720-5.

[15] Haacke EM, DelProposto ZS, Chaturvedi S, Sehgal V, Tenzer M, Neelavalli J, et al. Imaging cerebral amyloid angiopathy with susceptibility-weighted imaging. AJNR Am J Neuroradiol 2007;28:316-7.

[16] Wagle WA, Smith TW, Weiner M. Intracerebral haemorrhage caused by cerebral amyloid angiopathy: radiographic-pathological correlation. AJNR Am J Neuroradiol 1984;5:171-6.

[17] Miller JH, Wardlaw JM, Lammie GA. Intracerebral haemorrhage and cerebral amyloid angiopathy: CT features with pathological correlation. Clin Radiol 1999;54:422-9.

[18] Linn J, Herms J, Dichgans M, Brückmann H, Fesl G, Freilinger T, et al. Subarachnoid hemosiderosis and superficial cortical hemosiderosis in cerebral amyloid angiopathy. AJNR Am J Neuroradiol 2008;29:184-6.

[19] Fazekas F, Chawluk JB, Alavi A, Hurtig HI, Zimmerman RA. MR signal abnormalities at 1. 5T in Alzheimer's dementia and normal aging. AJNR Am J Neuroradiol 1987;8:421-6.

[20] Loes DJ, Biller J, Yuh WT, Hart MN, Godersky JC, Adams HP, et al. Leukoencephalopathy in cerebral amyloid angiopathy: MR imaging in four cases. AJNR Am J Neuroradiol 1990;11:485-8.

[21] Chung KK, Anderson NE, Hutchinson D, Synek B, Barber PA. Cerebral amyloid angiopathy related inflammation: three case reports and a review. J Neurol Neurosurg Psychiatry 2011;82:20-6.

[22] Greenberg SM, Finklestein SP, Schaefer PW. Petechial hemorrhages accompanying lobar hemorrhage: detection by gradient-echo MRI. Neurology 1996;46:1751-4.

[23] Izumihara A, Ishihara T, Iwamoto N, Yamashita K, Ito H. Postoperative outcome of 37 patients with lobar intracerebral hemorrhage related to cerebral amyloid angiopathy. Stroke 1999;30:29-33.

[24] Kloppenborg RP, Richard E, Sprengers ME, Troost D, Eikelenboom P, Nederkoorn PJ. Steroid responsive encephalopathy in cerebral amyloid angiopathy: a case report and review of evidence for immunosuppressive treatment. J Neuroinflammation 2010;7:18.

[25] Greenberg SM, Rosand J, Schneider AT, Creed Pettigrew L, Gandy SE, Rovner B, et al. A phase 2 study of tramiprosate for cerebral amyloid angiopathy. Alzheimer Dis Assoc Disord 2006;20:269-74.

Chapter 5

Genetics of Sporadic Cerebral Amyloid Angiopathy

Kristiina Rannikmäe and Cathie Sudlow

Additional information is available at the end of the chapter

http://dx.doi.org/10.5772/58870

1. Introduction

Cerebral amyloid angiopathies (CAA) can be divided into sporadic and hereditary forms. This chapter is focused on the genetics of sporadic CAA, but will first consider hereditary forms in brief.

1.1. Hereditary CAA

Amyloid-β protein (Aβ), the commonest amyloid subunit implicated in sporadic forms of CAA, is also involved in certain hereditary forms. Several other proteins are also associated with rare familial diseases in which CAA is a characteristic morphological feature. [1] Missense mutations within or just outside the Aβ peptide coding region of the APP gene result in clinicopathological phenotypes of early onset Alzheimer's disease (AD) and are associated with a neuropathological phenotype which includes prominent CAA – for example hereditary cerebral haemorrhage with amyloidosis of Dutch type (HCHWA-D), or with Italian, Arctic, Iowa, Piedmont and Flemish mutations. Severe Aβ CAA has also been well documented in cases of familial AD due to mutations in the presenilin (PSEN1 and PSEN2) genes. Familial CAAs associated with other proteins include BRI2 gene-related dementias (familial British dementia and familial Danish dementia), cystatin C gene mutations in hereditary cerebral haemorrhage with amyloidosis of Icelandic type, TTR gene mutations in meningo-vascular amyloidosis, hereditary prion disease with premature stop codon mutations and mutated gelsolin gene in familial amyloidosis of Finnish type. [1]

1.2. Sporadic CAA

Sporadic cerebral amyloid angiopathy is characterised by deposition of Aβ in leptomeningeal and cortical blood vessels. It has a prevalence in population-based autopsy studies of 20-40%

in non-demented and 50-60% in demented elderly people. [2] Neuropathological case-control and cross-sectional studies, as well as the increased incidence of intracerebral haemorrhage (ICH) in patients with Alzheimer's disease, suggest that CAA causes lobar ICH. [3, 4] CAA is also associated with increasing age, dementia, lobar brain microbleeds, leukoaraiosis, small cortical infarcts and superficial siderosis. [3, 5-7]

It is unknown why only a few people with CAA pathology develop an ICH, but it seems likely to involve biological pathways additional to and distinct from those involved in vascular amyloid deposition. Cases of CAA with ICH not only have a greater proportion of amyloid-laden blood vessels, [8] but also more often demonstrate severe CAA with associated vasculopathy. [8-11]

Identifying genetic polymorphisms associated with the presence of histopathologically confirmed CAA in general, as well as with the severe form of CAA thought to cause vessel rupture and ICH, should increase our understanding of the mechanisms leading to CAA and associated diseases, including CAA associated ICH.

Polymorphisms in the apolipoprotein E gene (APOE) [12] are associated with ICH as well as with other conditions in which CAA may be involved, including subarachnoid haemorrhage, lobar brain microbleeds, and AD. [7, 13-18] In vitro studies have shown that APOE influences Aβ conformation, fibril formation and toxicity, [19, 20] while in vivo mouse studies have confirmed a critical role for apolipoprotein E in Aβ deposition, toxicity and possibly clearance. [21, 22] It therefore seems likely that APOE influences risk of developing histopathologically confirmed, sporadic CAA. Other genetic polymorphisms are also likely to contribute to development of sporadic CAA.

Below we present and summarise the evidence for associations of polymorphisms in APOE or any other gene with histopathologically confirmed, sporadic CAA in adult humans. We then go on to consider the evidence for associations of APOE with the severe form of CAA. The evidence presented is based mainly on our two recently published systematic reviews of all relevant published studies, both of which incorporated a comprehensive search strategy, a thorough assessment of study quality, a series of meta-analyses, and an evaluation of the robustness of any positive findings to small study and other methodological biases. [23, 24] Figure 1 summarises the strategy used for identifying relevant studies and the numbers of studies (and study participants) identified.

2. Genetic associations with histopathologically-confirmed, sporadic CAA

While robust, large-scale evidence exists for an association of APOE with ICH attributed to CAA on the basis of clinical criteria, [15] studies assessing association of APOE with histopathologically confirmed CAA have had various methodological shortcomings (including small size), and reported results vary. Our systematic review of genetic associations with histopathologically confirmed, sporadic CAA sought all studies in which participants had been

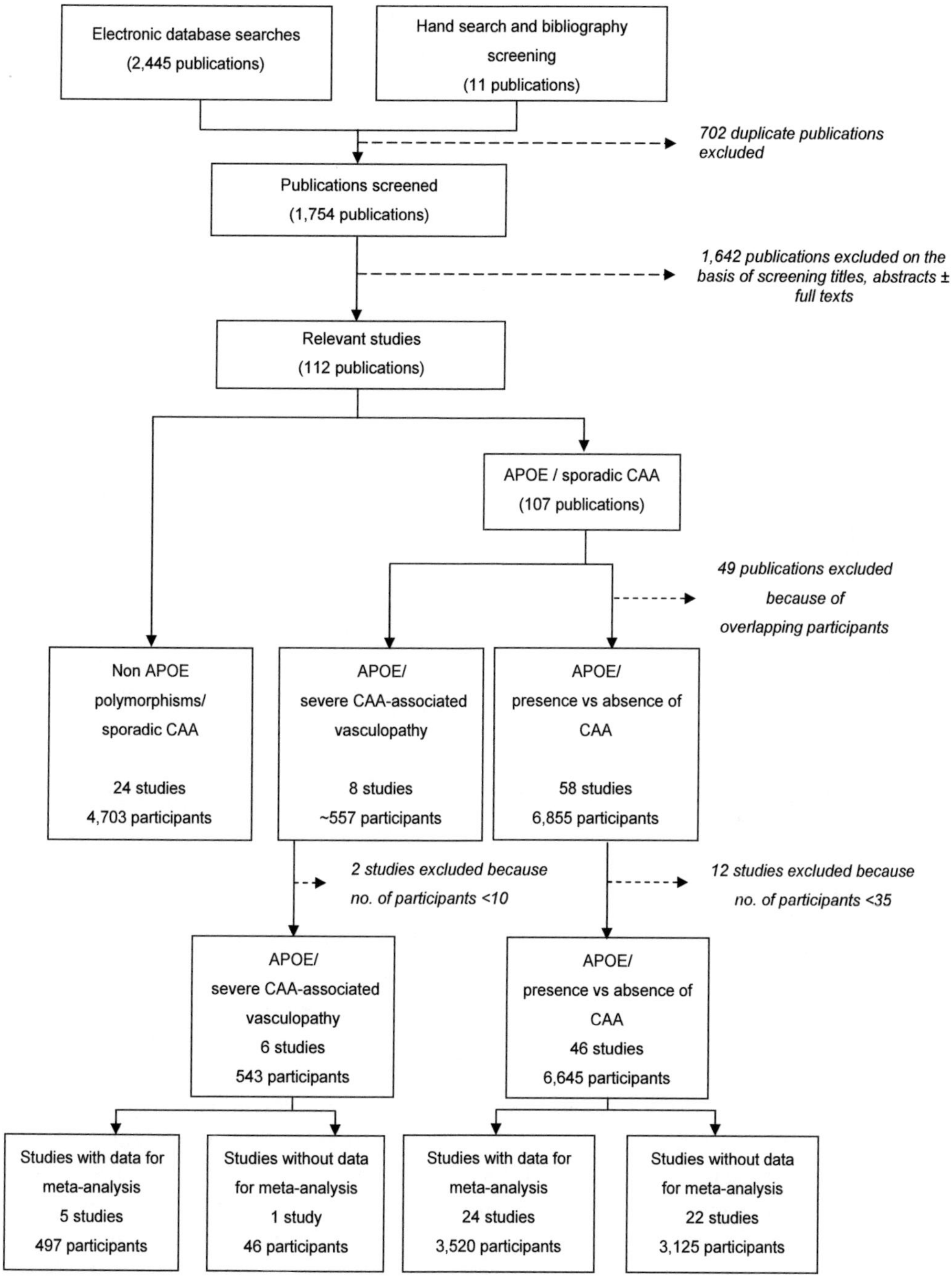

Figure 1. Selection of included studies

genotyped for any genetic polymorphism and had CAA assessed pathologically (using autopsy or biopsy). [23] Studies that had assessed genetic associations with CAA-associated ICH (CAAH) versus CAA-free controls were excluded, because these would not be able to distinguish a genetic association with the presence of CAA histopathology from an association with ICH.

2.1. APOE ε2/ε3/ε4 polymorphism and sporadic CAA

We identified 46 studies including 6645 participants with data about the APOE ε2/ε3/ε4 polymorphism and sporadic CAA (Figure 1). [25-70] These studies had used autopsy brains from clinical autopsy collections, a brain bank or a population-based prospective study. Participants' mean age was 70 to 85 years in most studies and about half were male. Almost 90% of participants were of European ancestry while around 10% were from Asian populations (all Japanese). About 30% of participants had clinical dementia (mainly AD), about 10% were known not to be demented and dementia status was not specified for the remainder. There was substantial variation in overall study quality. Genotyping reporting quality [71, 72] was generally limited and methods for pathological assessment were very variable. Larger studies tended to be of higher quality. [23]

2.1.1. APOE ε4 and CAA

Meta-analyses were possible of data from just over half of these studies (including just over half of the participants), and showed a significant association between ε4+ genotypes and presence of CAA (OR 2.67, 95% CI 2.31 to 3.08), although there was significant heterogeneity between the studies' results (Figure 2). There were no significant differences between subgroups of studies based on dementia status, ethnicity or overall study quality score (Figures 2 and 3). Six studies (443 participants) made only a qualitative statement, [27, 30, 33, 37, 40, 42] reporting either no significant association or a trend towards association with APOE ε4, while 16 studies (2682 participants) provided no information about the association. [25, 28, 29, 31, 32, 34-36, 38, 39, 41, 43, 46, 47, 50, 57]

Failsafe N calculations [73] showed that a null study of >137, 000 participants would be required to bring the association of ε4+ genotypes with CAA from the meta-analysis to a just statistically non-significant level. This makes it unlikely that there might plausibly be enough participants in unpublished, unreported or otherwise unretrieved null studies to make this significant result non-significant, and suggests that the association of APOE ε4+ genotypes with histopathologcally confirmed, sporadic CAA is real and robust. Meta-analysis of the association of APOE ε4 allele dose with CAA among 12 studies (1706 participants) providing quantitative data showed a significant increase in the odds of having CAA with increasing dose of the ε4 allele (Figure 4). Two further studies (117 participants) provided a qualitative statement about the association supported this result. [40, 64] Failsafe N calculations showed that it would require a null study of >7000 participants to bring the stronger association with CAA of ε4 homozygous versus heterozygous genotypes to a just non-significant level, suggesting that the finding of a dose-response relationship between APOE ε4 and CAA is real and robust.

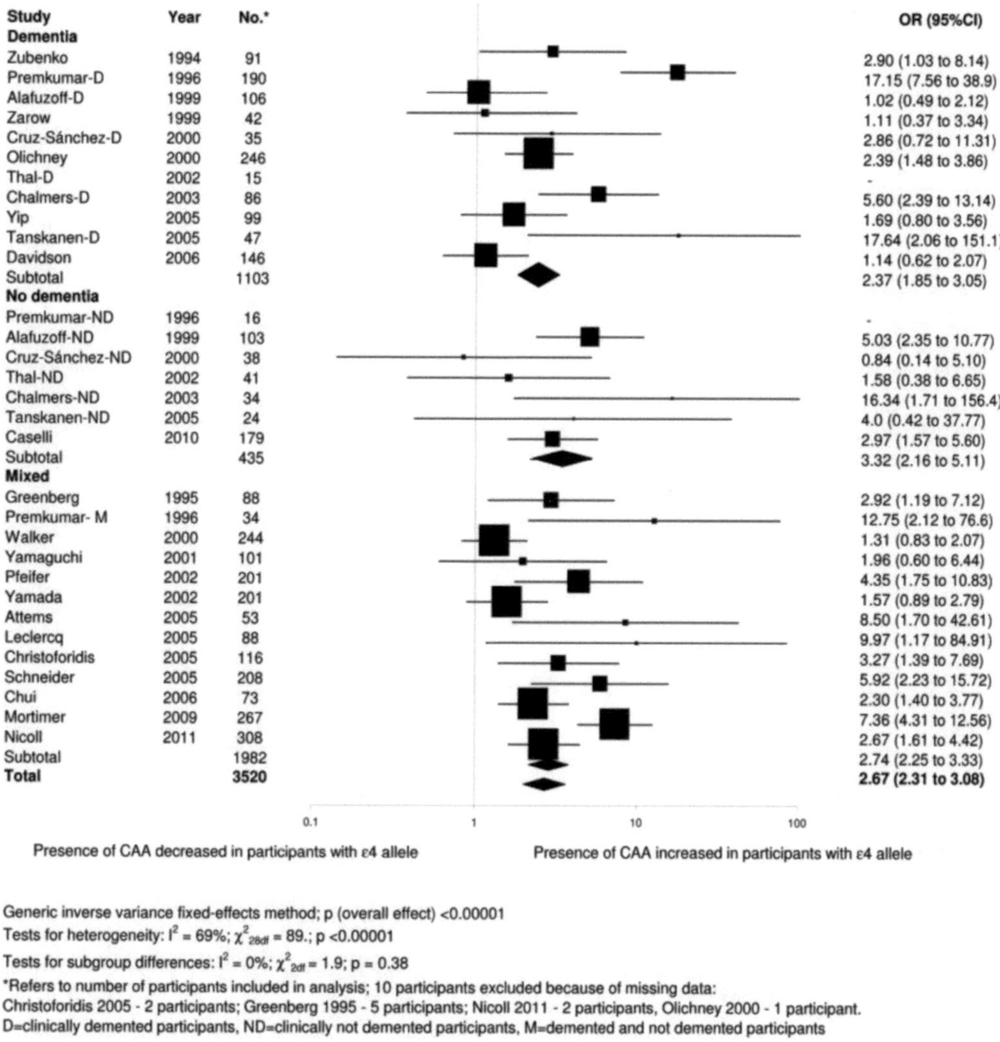

The squares represent study-specific odds ratios (ORs), with their size proportional to their statistical weight by the generic inverse variance method. Horizontal lines represent 95% confidence intervals (CIs). Diamonds represent pooled ORs, and their width represents the 95% CI. Higher score represents better study quality.
Reproduced from "Genetics of cerebral amyloid angiopathy: systematic review and meta-analysis" Rannikmäe K, Samarasekera N, Martĩnez-Gonzãlez NA, Al-Shahi Salman R, Sudlow C. Journal of Neurology Neurosurgery and Psychiatry 2013;84(901-8), with permission from BMJ Publishing Group Ltd.

Figure 2. Meta-analysis of association of APOE ε4+ vs ε4- genotypes with CAA by participants' dementia status

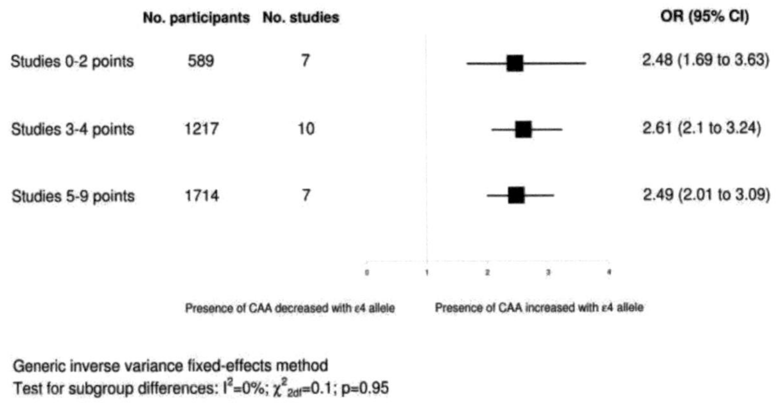

The squares represent pooled ORs and their width represents the 95% CI. Higher score represents better study quality. Reproduced from "Genetics of cerebral amyloid angiopathy: systematic review and meta-analysis" Rannikmäe K, Samarasekera N, Martīnez-Gonzālez NA, Al-Shahi Salman R, Sudlow C. Journal of Neurology Neurosurgery and Psychiatry 2013;84(901-8), with permission from BMJ Publishing Group Ltd.

Figure 3. Subgroup analysis based on study quality scores.

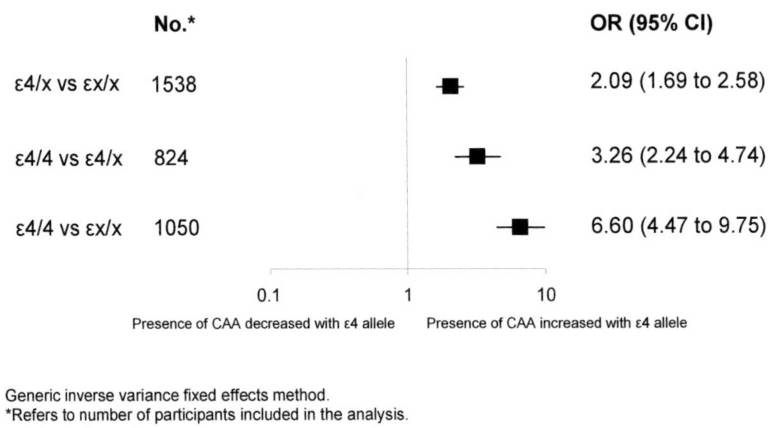

The squares represent pooled ORs and their width represents the 95% CI.

Figure 4. Meta-analysis of effects of APOE ε4 dose (ε4--/ε4+-/ε4++genotypes) on presence vs absence of CAA

2.1.2. APOE ε2 and CAA

Meta-analysis of the association of APOE ε2+ versus ε2- genotypes with CAA among 11 studies (1640 participants) showed borderline significant decreased odds of CAA with APOE ε2+ genotypes (OR 0.73, 95% CI 0.53 to 1.00, p=0.05). Two studies (213 participants) provided a qualitative statement; neither reported a significant association. [60, 64]

2.1.3. Summary and discussion

There is, therefore, robust evidence for a highly significant, dose-dependent association between APOE ε4 and pathologically proven CAA, which does not vary significantly with dementia status, ethnicity, or study quality. However, there is no clear overall association between APOE ε2 and histopathologically confirmed CAA. Lack of variation in the effect of APOE ε4 by study size and the very large failsafe N showed that this association could not plausibly be explained by publication, reporting or any other small study bias. It is important to note that the quality of studies included in our systematic review was generally limited when assessed against current reporting standards. [71, 72] However, there were - reassuringly - no significant subgroup differences by study quality score.

The prevalence of CAA in Alzheimer's disease is over 70% but the relationship between CAA and AD is still poorly understood. Although the diagnostic criteria for dementia and the participant inclusion criteria varied between the studies in our systematic review (some excluding cases with severe dementia), the demonstration of a similar association in those with and without clinical dementia suggests that the association of APOE ε4 with CAA is independent of its known association with dementia (mainly Alzheimer's disease).

Pathological assessment in the included studies was very variable. Indeed, there is no widely accepted, standardized histopathological grading system for CAA, [74] and no comparative studies to determine the most accurate method for assessing CAA (although the suggested method is a combination of Thioflavin S/T or Congo Red with immunohistochemistry). [75] CAA assessment location also varied widely, possibly influencing the rate of CAA detection, since a greater burden of CAA is generally reported in the occipital or parietal lobes, albeit with a higher frequency of frontal lobe involvement reported in studies from China and Japan. [74] This is important because genetic associations may differ by CAA location and subtype. For example, there is preliminary evidence that APOEε4 may be associated with CAA type 1 (where CAA is found in cortical capillaries), and ε2 with CAA type 2 (where amyloid is deposited in leptomeningeal and cortical vessels with the exception of cortical capillaries). [26] In addition, since APOE effects on ICH may vary with ethnicity, there may also be ethnic variation in genetic associations with CAA, but these have not yet been widely enough studied in non-white populations to assess this reliably. [76]

2.2. Associations between other genetic polymorphisms and sporadic CAA

In our systematic review, few polymorphisms other than APOE had been studied in more than a few hundred participants or in more than one study and there were not enough data for meta-analysis (Figure 1, Table 1). [39, 46, 50, 61, 63, 77-95] Thus, there were too few studies and participants to draw firm conclusions about the effect of other genetic polymorphisms. However, there were some suggestive positive associations with CAA. First, there was a consistent trend towards an association with CAA of a single nucleotoide polymorphism (SNP) in the transforming growth factor-β1 (TGF-β1) gene in two studies (449 participants). [82, 85] If real, this may occur through an influence of TGF-β1 on Aβ clearance and deposition through activation of astrocytes and microglia. Second, there were significant associations in one study (723 participants) of SNPs in the translocase of outer mitochondrial membrane 40 (TOMM40) gene with vascular amyloid burden but not with ICH attributed to CAA, [88] which could be through interaction of TOMM40 with APOE ε2 or through its effects on Aβ mitochondrial transport. Finally, one study (544 participants) found an association of a SNP in the complement component receptor 1 (CR1) gene with both CAA severity and ICH attributed to CAA, possibly occurring via altered clearance of Aβ peptide. [96] Other studies found no overall significant associations, although some reported associations in particular subgroups (Table 1).

Gene	Location / Polymorphism	No. of studies	No. of participants	Summary of results
TGF-β1	rs1800470	2	449	Consistent trend for positive association between T allele and CAA
TOMM40	rs2075650, rs34404554, rs11556505, rs769449, rs12972156, rs12972970, rs157582, rs184017, rs157581, rs283815, rs157580, rs439401, rs34095326, rs10119	1	723	SNPs associated with vascular amyloid burden
CR1 gene	rs6656401	1	544	Associated with severity of CAA pathology
LRP1 (low-density lipoprotein receptor 1)	rs1799986	3	597	No overall significant associations (inconsistent trends and in some cases associations in subgroups)
ACT (α1 antichymotrypsin)	signal region of the gene → A/T alleles that determine the aminoacid alanine or threonine**	2	235	
CYP46	rs754203	2	524	
ACE (angiotensin 1 converting enzyme)	intron 16 insertion/deletion of a 287 bp sequence	2	239	
Gene				
PS1 (presenilin-1);BCHE (butyrylcholinesterase);DXS1047 locus;APOE promoter;A2M (α2 macroglobulin);PON1 (paraoxonase);NEP (neprilysin);OLR1 (oxidized low-density lipoprotein receptor 1);LRP (low density lipoprotein receptor related protein);CYP46; CH25H*1;VEGF (vascular endothelial growth factor);IL-1A;IL-1B;IL-33;GSTO1-1 (glutathione S-transferase omega-1);SORL1 (sortilin related receptor);CTSD (cathepsin D); AβPP and AβPppromoter;		18	50-380*	

*Range of participant numbers in individual studies **probably rs4943

Adapted by permission from BMJ Publishing Group Limited, from "Genetics of cerebral amyloid angiopathy: systematic review and meta-analysis" Rannikmäe K, Samarasekera N, Martínez-González NA, Al-Shahi Salman R, Sudlow C. Journal of Neurology Neurosurgery and Psychiatry 2013;84(901-8)

Table 1. Summary of studies of non-APOE polymorphisms and CAA

3. APOE allele-specific associations with severe CAA-associated vasculopathy

The systematic review and series of meta-analyses presented in the previous section confirmed an association between histopathologically diagnosed CAA and APOE ε4, but not APOE ε2. However a recent large scale genetic association study found that both ε2 and ε4 containing genotypes were associated with clinically diagnosed CAA, manifesting as lobar ICH attributed to CAA. [15] Furthermore, APOE ε2 has been found to predict initial haematoma volume, haematoma expansion, increased mortality and poor functional outcome after lobar ICH. [17, 97] The currently favoured popular explanation for these findings is, that APOE ε4 enhances deposition of amyloid-β in cerebral blood vessel walls, while ε2 promotes haemorrhage from amyloid-laden blood vessels by increasing specific CAA-related vasculopathic changes (Figure 5). [8, 25, 98]

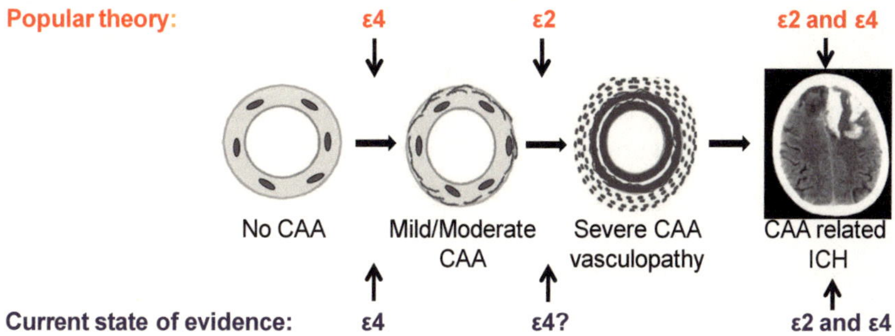

Adapted from Figure 1 in Acta Neuropathologica 2005;110: 345–359 "Sporadic cerebral amyloid angiopathy: pathology, clinical implications, and possible pathomechanisms", Johannes Attems, with kind permission from Springer Science and Business Media and Professor Attems.

Figure 5. Proposed theory and current state of evidence about associations between APOE and CAA phenotype

In a further recent systematic review, we reviewed the evidence for this hypothesis. [24] The main focus of this work was on assessing the potential influence of APOE genotypes on severe CAA preceding rupture. To avoid selection bias, we excluded studies with participants selected on the basis of having had a CAA-related ICH, since APOE ε2 and ε4 are already known to be associated with this phenotype, and severe CAA is commoner in such cases. This review sought all studies, which had conducted both APOE genotyping and histopathological assessment for CAA, including assessment for severe CAA with associated vasculopathic changes (blood vessel dilatation, microaneurysm formation, fibrinoid degeneration, cracking and double-barrelling of the vessel wall, and paravascular leakage of blood). The assessment for this severe form of CAA could have occurred either as part of the Vonsattel grading scale, [9] which includes such changes in its 'severe' category or through specifically reporting on some or all of the relevant histopathological characteristics. From 1754 publications screened, we identified six eligible studies, which included 543 eligible participants (Figure 1). [8, 25, 62, 64, 99, 100] Only one of the six studies had previously reported on the association between the APOE genotype and severe CAA (assessed using Vonsattel scale), finding a significantly greater frequency of APOE ε2 in severe versus moderate CAA cases. [25] This study and four others that had rated CAA on the Vonsattel scale, between them including 497 eligible participants (92% of all 543 potentially eligible participants), [62, 64, 99, 100] were able to share their unpublished data for collaborative meta-analyses.

The five studies included in the meta-analyses used autopsy brains either from brain tissue banks or from a population-based prospective study with an autopsy component. There were 57 to 227 eligible participants per study, mean age at death was 78 to 84 years and about half of all participants were male. Three studies (357 participants) were conducted in predominantly white populations in the USA while information on ethnicity was unavailable for two studies (140 participants). About 50% of participants had clinical dementia (mainly neuropathologically confirmed AD), about 20% were known not to be demented and in the remaining 30% dementia status was unknown. The quality of genotyping and of pathology assessment was generally very good when assessed against current reporting standards. [71, 72] As in the first of our two systematic reviews, methods for pathological assessment were variable, reflecting a lack of agreed standards for CAA pathology assessment at the time these studies were conducted. [24]

Among the 353 individuals in these five studies who had CAA present on histopathological assessment, meta-analyses found a significant association of ε4+ versus ε4- genotypes with severe versus mild/moderate CAA (OR 2.5, 95% CI 1.4 to 4.5, p=0.002) but no significant association with severe versus moderate CAA (OR 1.7, 95% CI 0.9 to 3.1, p=0.11) (Figure 6). There was no significant heterogeneity between individual studies' results. For ε2+ versus ε2- genotypes, the associations with severe CAA versus mild/moderate CAA and with severe versus moderate CAA were non-significant, with wide confidence intervals due to small numbers of participants, particularly in the ε2+ group, which included 22 individuals, only seven of whom had severe CAA (Figure 7). There was moderate heterogeneity between individual studies' results for severe versus mild/moderate CAA (I^2=52%; χ^2_{3df}=6.2; p=0.1) and minimal heterogeneity for severe versus moderate CAA (I^2=11%; χ^2_{3df}=3.4; p=0.3). Results were similar and conclusions unchanged for the ε4+ and ε2+ genotypes when ε3ε3 genotypes were used as the comparison group (rather than ε4- or ε2-). Associations with the presence versus absence of CAA were consistent with results from the previous published systematic review [23], showing a clearly significant association with ε4+ (ε4+ versus ε4-: OR 4.8, 95% CI 3.0 to 7.6, p<0.00001) but not with ε2+ genotypes (ε2+ versus ε2-: OR 0.38, 95% CI 0.1 to 1.0, p=0.05).

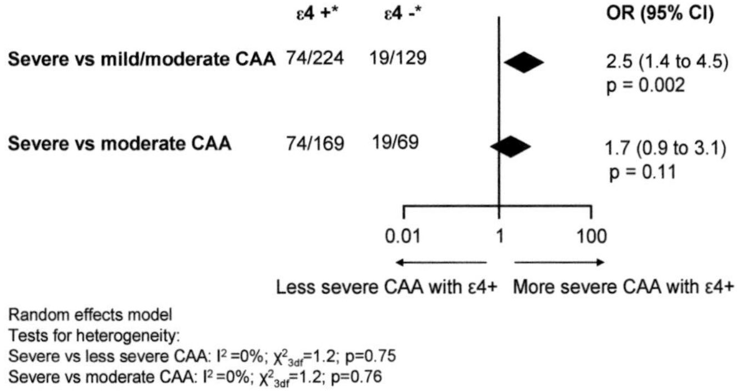

The diamonds represent pooled OR across studies and the width of the diamonds represents 95% confidence intervals (CIs).

*ε4+: number of subjects with an ε4 allele and severe CAA / total number of subjects with an ε4 allele and any severity of CAA *or* severe/moderate CAA

ε4-: number of subjects without an ε4 allele and severe CAA / total number of subjects without an ε4 allele and any severity of CAA *or* severe/moderate CAA

Results of two studies conducted in one centre were combined for the analyses. [25, 64]

Adapted by permission from BMJ Publishing Group Limited, from "APOE associations with severe CAA-associated vasculpathic changes: collaborative meta-analysis" Rannikmäe K, Kalaria RN, Greenberg SM, Chui HC, Schmitt FA, Samarasekera N, et al. Journal of Neurology Neurosurgery and Psychiatry 2014;85(3):300-5.

Figure 6. Meta-analysis of APOE-ε4 associations with severe CAA

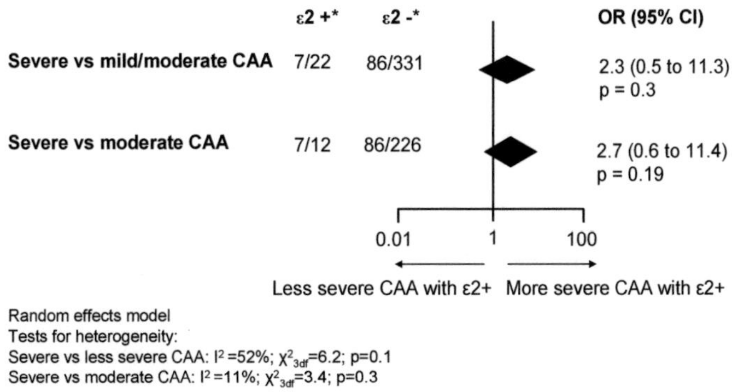

The diamonds represent pooled OR across studies and the width of the diamonds represents 95% confidence intervals (CIs). *ε2+: Number of subjects with an ε2 allele and severe CAA / total number of subjects with an ε2 allele and any severity of CAA *or* severe/moderate CAA ε2-: number of subjects without an ε2 allele and severe CAA / total number of subjects without an ε2 allele and any severity of CAA *or* severe/moderate CAA. Results of two studies conducted in one centre were combined for the analyses. [25, 64]

Adapted by permission from BMJ Publishing Group Limited, from "APOE associations with severe CAA-associated vasculpathic changes: collaborative meta-analysis." Rannikmäe K, Kalaria RN, Greenberg SM, Chui HC, Schmitt FA, Samarasekera N, et al. Journal of Neurology Neurosurgery and Psychiatry 2014;85(3):300-5.

Figure 7. Meta-analysis of APOE ε2 associations with severe CAA

Thus, contrary to what has been suggested, [8, 25] a systematic assessment of the relevant evidence suggests a possible association of APOE ε4 but not of APOE ε2 with progression to severe CAA (Figure 5). However, this does not exclude a biologically meaningful association with APOE ε2 since, despite including data from almost all relevant cases from the published literature, total numbers of individuals included in our meta-analyses were relatively small and confidence intervals wide, especially for analyses of the effects of APOE ε2. There were other limitations too. First, methods for histopathological assessment varied between studies, potentially introducing heterogeneity and reducing the likelihood of detecting a consistent effect across studies. Second, APOE allele-specific effects on severe CAA may differ according to the presence or absence of Alzheimer's disease, particularly for APOE ε2, which has been associated with a decreased risk of Alzheimer's dementia. [101] Informative subgroup analysis to explore potential causes of heterogeneity could not be performed, however, because of the small overall numbers of participants and because dementia status was unknown for a large number of participants. Third, while the studies included assessed those severe CAA-associated vasculopathic changes that are specifically alluded to in the Vonsattel scale, other vasculopathic changes may also be relevant. Fourth, both APOE allele-specific and other genetic associations may differ by CAA subtype. The preliminary evidence that APOE ε4 may be associated with CAA type 1 (where CAA is found in cortical capillaries), and ε2 with CAA type 2 (where amyloid is deposited in leptomeningeal and cortical vessels with the exception of cortical capillaries) [26] suggests that CAA types 1 and 2 may represent different pathological entities, and – if so-the mechanisms and genetic risk factors for severe CAA and ICH could also differ. Finally, there may be other genetic influences that interact with APOE ε2 to increase risk of or protect against severe CAA and ICH.

4. Conclusions

There is strong evidence that APOE ε4 promotes cerebral amyloid angiopathy, and further evidence to suggest that ε4 may increase the risk of developing severe CAA among those with CAA. However, there is not convincing evidence to support the theory that APOE ε2 promotes progression to severe CAA-related vasculopathic changes so leading to vessel rupture and ICH. Much larger numbers of individuals will need to be included in CAA histopathology studies if reliable conclusions are to be drawn about the specific effects of APOE ε2, while bearing in mind that APOE genotype will not be the only genetic influence on CAA. Future research efforts in this area will also be helped substantially by the development and use of an internationally-agreed, standardised histopathological grading system for CAA (including assessment of CAA types 1 and 2), and by the consistent reporting of dementia – and specifically Alzheimer's disease – status [102] among individuals included in CAA histopathology studies.

http://dx.doi.org/10.5772/58870

Acknowledgements

The authors would like to acknowledge for their scientific input Professor Rustam Al-Shahi Salman, Dr Neshika Samarasekera and Nahara Anani Martînez-Gonzâlez (Centre for Clinical Brain Sciences, University of Edinburgh, Edinburgh, UK), Professor Rajesh N Kalaria (Institute for Ageing and Health, Newcastle University, Newcastle upon Tyne, UK), Professor Steven M Greenberg (Department of Neurology, Stroke Research Centre, Massachusetts General Hospital, Boston, Massachusetts, USA), Professor Helena C Chui (Department of Neurology, University of Southern California, Los Angeles, California, USA) and Professor Frederick A Schmitt (Department of Neurology, Sanders-Brown Center on Aging, University of Kentucky, Lexington, Kentucky, USA).

Author details

Kristiina Rannikmäe[1] and Cathie Sudlow[1,2]

*Address all correspondence to: cathie.sudlow@ed.ac.uk

1 Centre for Clinical Brain Sciences, University of Edinburgh, Edinburgh, UK

2 Institute for Genetics and Molecular Medicine, University of Edinburgh, Edinburgh, UK

References

[1] Revesz T, Holton JL, Lashley T, Plant G, Frangione B, Ristagno A et al. Genetics and molecular pathogenesis of sporadic and hereditary cerebral amyloid angiopathies. Acta Neuropatholologica 2009;118:115–30.

[2] Charidiomou A, Gang Q, Werring DJ. Sporadic cerebral amyloid angiopathy revisited: recent insights into pathophysiology and clinical spectrum. Journal of Neurology Neurosurgery and Psychiatry 2012;83(2)124-37.

[3] Samarasekera N, Smith C, Al-Shahi Salman R. The association between cerebral amyloid angiopathy and intracerebral haemorrhage: systematic review and meta-analysis. Journal of Neurology Neurosurgery and Psychiatry 2012;83(3)275-81.

[4] Chi NF, Chien LN, Ku HL, Hu CJ, Chiou HY. Alzheimer disease and risk of stroke: a population-based cohort study. Neurology 2013;80(8)705-11.

[5] Masuda J, Tanaka K, Ueda K, Omae T. Autopsy study of incidence and distribution of cerebral amyloid angiopathy in Hisayama, Japan. Stroke 1988;19(2)205-10.

[6] Neuropathology Group.Medical Research Council Cognitive Function and Aging Study. Pathological correlates of late-onset dementia in a multicentre, community

based population in England and Wales. Neuropathology Group of the Medical Research Council Cognitive Function and Ageing Study (MRC CFAS). Lancet 2001;357(9251)169-75.

[7] Linn J, Halpin A, Demaerel P, Ruhland J, Giese AD, Dichgans M, et al. Prevalence of superficial siderosis in patients with cerebral amyloid angiopathy. Neurology 2010;74(17)1346-50.

[8] McCarron MO, Nicoll JA, Stewart J, Ironside JW, Mann DM, Love S, et al. The apolipoprotein E epsilon2 allele and the pathological features in cerebral amyloid angiopathy-related hemorrhage. Journal of Neuropathology and Experimental Neurology 1999;58(7)711-8.

[9] Vonsattel JPG, Myers RH, Hedley-Whyte ET, Ropper AH, Bird ED, Richardson EP Jr. Cerebral amyloid angiopathy without and with cerebral hemorrhages: a comparative histological study. Annals of Neurology 1991;30(5)637-49.

[10] Okazaki H, Reagan TJ, Campbell RJ. Clinicopathological studies of primary cerebral amyloid angiopathy. Mayo Clinic Proceedings 1979;54(1)22-31.

[11] Mandybur TI. Cerebral amyloid angiopathy: the vascular pathology and complications. Journal of Neuropathology and Experimental Neurology 1986;45(1)79-90.

[12] Eichner JE, Dunn ST, Perveen G, Thompson DM, Stewart KE, Stroehla BC. Apolipoprotein E polymorphism and cardiovascular disease: a HuGE review. American Journal of Epidemiology 2002;155(6)487-95.

[13] Paternoster L, Chen W, Sudlow CLM. Genetic determinants of white matter hyperintensities on brain scans: a systematic assessment of 19 candidate gene polymorphisms in 46 studies in 19, 000 subjects. Stroke 2009;40(6)2020-6.

[14] Sudlow C, Martķnez Gonzįlez NA, Kim J, Clark C. Does Apolipoprotein E genotype influence the risk of ischemic stroke, intracerebral hemorrhage, or subarachnoid hemorrhage? Systematic review and meta-analyses of 31 studies among 5961 cases and 17, 965 controls. Stroke 2006;37(2)364-70.

[15] Biffi A, Sonni A, Anderson CD, Kissela B, Jagiella JM, Schmidt H, et al. Variants of APOE influence risk of deep and lobar intracerebral hemorrhage. Annals of Neurology 2010;68(6)934-43.

[16] Maxwell SS, Jackson CA, Paternoster L, Cordonnier C, Thijs V, Al-Shahi Salman R, et al. Genetic associations with brain microbleeds: Systematic review and meta-analyses. Neurology 2011;77(2)158-67.

[17] Biffi A, Anderson CD, Jagiella JM, Schmidt H, Kissela B, Hansen BM, et al. APOE genotype and extent of bleeding and outcome in lobar intracerebral haemorrhage: a genetic association study. Lancet Neurology 2011;10(8)702-9.

[18] Bertram L, McQueen MB, Mullin K, Blacker D, Tanzi RE. Systematic meta-analyses of Alzheimer disease genetic association studies: the AlzGene database. Nature Genetics 2007;39(1)17-23.

[19] Lambert MP, Barlow AK, Chromy BA, Edwards C, Freed R, Liosatos M, et al. Diffusible, nonfibrillar ligands derived from Abeta1-42 are potent central nervous system neurotoxins. Proceedings of the National Academy of Sciences of the United States of America 1998;95(11)6448-53.

[20] Wood SJ, Chan W, Wetzel R. Seeding of A beta fibril formation is inhibited by all three isotypes of apolipoprotein E. Biochemistry 1996;35(38)12623-8.

[21] Holtzman DM, Bales KR, Wu S, Bhat P, Parsadanian M, Fagan AM, et al. Expression of human apolipoprotein E reduces amyloid-beta deposition in a mouse model of Alzheimer's disease. The Journal of Clinical Investigation 1999;103(6)R15-R21.

[22] Holtzman DM, Bales KR, Tenkova T, Fagan AM, Parsadanian M, Sartorius LJ, et al. Apolipoprotein E isoform-dependent amyloid deposition and neuritic degeneration in a mouse model of Alzheimer's disease. Proceedings of the National Academy of Sciences of the United States of America 2000;97(6)2892-7.

[23] Rannikmäe K, Samarasekera N, Martīnez-Gonzālez NA, Al-Shahi Salman R, Sudlow C. Genetics of cerebral amyloid angiopathy: systematic review and meta-analysis. Journal of Neurology Neurosurgery and Psychiatry 2013;84(8)901-8.

[24] Rannikmäe K, Kalaria RN, Greenberg SM, Chui HC, Schmitt FA, Samarasekera N, et al. APOE associations with severe CAA-associated vasculpathic changes: collaborative meta-analysis. Journal of Neurology Neurosurgery and Psychiatry 2014;85(3): 300-5.

[25] Greenberg SM, Vonsattel JP, Segal AZ, Chiu RI, Clatworthy AE, Liao A, et al. Association of apolipoprotein E epsilon2 and vasculopathy in cerebral amyloid angiopathy. Neurology 1998;50(4)961-5.

[26] Thal DR, Ghebremedhin E, Rüb U, Yamaguchi H, Del Tredici K, Braak H. Two types of sporadic cerebral amyloid angiopathy. Journal of Neuropathology and Experimental Neurology 2002;61(3)282-93.

[27] Thomas AJ, Morris CM, Ferrier IN, Kalaria RN. Distribution of amyloid beta 42 in relation to the cerebral microvasculature in an elderly cohort with Alzheimer's disease. Annals of the New York Academy of Sciences 2000;903(83-8.

[28] Alafuzoff I, Aho L, Helisalmi S, Mannermaa A, Soininen H. Beta-amyloid deposition in brains of subjects with diabetes. Neuropathology and applied neurobiology 2009;35(1)60-8.

[29] Attems J, Lauda F, Jellinger KA. Unexpectedly low prevalence of intracerebral hemorrhages in sporadic cerebral amyloid angiopathy: An autopsy study. Journal of Neurology 2008;255(1)70-6.

[30] Berg L, McKeel DW Jr, Miller JP, Storandt M, Rubin EH, Morris JC, et al. Clinicopathologic studies in cognitively healthy aging and Alzheimer disease: Relation of histologic markers to dementia severity, age, sex, and apolipoprotein E genotype. Archives of Neurology 1998;55(3)326-35.

[31] Etiene D, Kraft J, Ganju N, Gomez-Isla T, Gemelli B, Hyman BT, et al. Cerebrovascular pathology contributes to the heterogeneity of Alzheimer's disease. Journal of Alzheimer's Disease 1998;1(2)119-34.

[32] Honig LS, Kukull W, Mayeux R. Atherosclerosis and AD: analysis of data from the US National Alzheimer's Coordinating Center. Neurology 2005;64(3)494-500.

[33] Jicha GA, Parisi JE, Dickson DW, Cha RH, Johnson KA, Smith GE, et al. Age and apoE associations with complex pathologic features in Alzheimer's disease. Journal of the Neurological Sciences 2008;273(1-2)34-9.

[34] Lashley T, Holton JL, Gray E, Kirkham K, O'Sullivan SS, Hilbig A, et al. Cortical alpha-synuclein load is associated with amyloid-beta plaque burden in a subset of Parkinson's disease patients. Acta neuropathologica 2008;115(4)417-25.

[35] Lewis H, Beher D, Cookson N, Oakley A, Piggott M, Morris CM, et al. Quantification of Alzheimer pathology in ageing and dementia: age-related accumulation of amyloid-beta(42) peptide in vascular dementia. Neuropathology and applied neurobiology 2006;32(2)103-18.

[36] Love S, Nicoll JA, Hughes A, Wilcock GK. APOE and cerebral amyloid angiopathy in the elderly. NeuroReport 2003;14(11)1535-6.

[37] Oyama F, Shimada H, Oyama R, Ihara Y. Apolipoprotein E genotype, Alzheimer's pathologies and related gene expression in the aged population. Molecular Brain Research 1995;29(1)92-8.

[38] Petrovitch H, Ross GW, He Q, Uyehara-Lock J, Markesbery W, Davis D, et al. Characterization of Japanese-American men with a single neocortical AD lesion type. Neurobiology of aging 2008;29(10)1448-55.

[39] Peuralinna T, Tanskanen M, Mäkela M, Polvikoski T, Paetau A, Kalimo H, et al. APOE and AbeetaPP gene variation in cortical and cerebrovascular amyloid-beeta pathology and Alzheimer's Disease: a population-based analysis. Journal of Alzheimer's Disease 2011;26(2)377-85.

[40] Roher AE, Kuo YM, Esh C, Knebel C, Weiss N, Kalback W, et al. Cortical and leptomeningeal cerebrovascular amyloid and white matter pathology in Alzheimer's disease. Molecular Medicine 2003;9(3-4)112-22.

[41] Sonnen JA, Larson EB, Walker RL, Haneuse S, Crane PK, Gray SL, et al. Nonsteroidal anti-inflammatory drugs are associated with increased neuritic plaques. Neurology 2010;75(13)1203-10.

[42] Stopa EG, Butala P, Salloway S, Johanson CE, Gonzalez L, Tavares R, et al. Cerebral cortical arteriolar angiopathy, vascular beta-amyloid, smooth muscle actin, Braak stage, and APOE genotype. Stroke 2008;39(3)814-21.

[43] Tiraboschi P, Sabbagh MN, Hansen LA, Salmon DP, Merdes A, Gamst A, et al. Alzheimer disease without neocortical neurofibrillary tangles: "A second look". Neurology 2004;62(7)1141-7.

[44] Caselli RJ, Walker D, Sue L, Sabbagh M, Beach T. Amyloid load in nondemented brains correlates with APOE e4. Neuroscience Letters 2010;473(3)168-71.

[45] Cruz-Sįnchez FF, Durany N, Thome J, Riederer P, Zambón D. Correlation between apolipoprotein-E polymorphism and Alzheimer's disease pathology. Journal of Alzheimer's Disease 2000;2(3-4)223-9.

[46] Durany N, Ravid R, Riederer P, Cruz-Sįnchez FF. Increased frequency of the alpha-1-antichymotrypsin T allele in cerebral amyloid angiopathy. Neuropathology 2000;20(3)184-9.

[47] Nelson PT, Abner EL, Schmitt FA, Kryscio RJ, Jicha GA, Smith CD, et al. Modeling the association between 43 different clinical and pathological variables and the severity of cognitive impairment in a large autopsy cohort of elderly persons. Brain Pathology 2010;20(1)66-79.

[48] Nicoll JA, Savva GM, Stewart J, Matthews FE, Brayne C, Ince P, et al. Association between APOE genotype, neuropathology and dementia in the older population of England and Wales. Neuropathology and applied neurobiology 2011;37(3)285-94.

[49] Olichney JM, Hansen LA, Hofstetter CR, Lee JH, Katzman R, Thal LJ. Association between severe cerebral amyloid angiopathy and cerebrovascular lesions in Alzheimer disease is not a spurious one attributable to apolipoprotein E4. Archives of Neurology 2000;57(6)869-74.

[50] Pahnke J, Walker LC, Schroeder E, Vogelgesang S, Stausske D, Walther R, et al. Cerebral beeta-amyloid deposition is augmented by the –491AA promoter polymorphism in non-demented elderly individuals bearing the apolipoprotein E e4 allele. Acta neuropathologica 2003;105(1)25-9.

[51] Pfeifer LA, White LR, Ross GW, Petrovitch H, Launer LJ. Cerebral amyloid angiopathy and cognitive function: The HAAS autopsy study. Neurology 2002;58(11)1629-34.

[52] Premkumar DR, Cohen DL, Hedera P, Friedland RP, Kalaria RN. Apolipoprotein E-epsilon4 alleles in cerebral amyloid angiopathy and cerebrovascular pathology associated with Alzheimer's disease. The American Journal of Pathology 1996;148(6)2083-95.

[53] Schneider JA, Bienias JL, Wilson RS, Berry-Kravis E, Evans DA, Bennett DA. The apolipoprotein E epsilon4 allele increases the odds of chronic cerebral infarction detected at autopsy in older persons. Stroke 2005;36(5)954-9.

[54] Walker LC, Pahnke J, Madauss M, Vogelgesang S, Pahnke A, Herbst EW, et al. Apolipoprotein E4 promotes the early deposition of Abeta42 and then Abeta40 in the elderly. Acta neuropathologica 2000;100(1)36-42.

[55] Yamada M. Risk factors for cerebral amyloid angiopathy in the elderly. Annals of the New York Academy of Sciences 2002;977(37-44.

[56] Yamaguchi H, Sugihara S, Ogawa A, Oshima N, Ihara Y. Alzheimer beta amyloid deposition enhanced by ApoE 4 gene precedes neurofibrillary pathology in the frontal association cortex of nondemented senior subjects. Journal of Neuropathology and Experimental Neurology 2001;60(7)731-9.

[57] Zipser BD, Johanson CE, Gonzalez L, Berzin TM, Tavares R, Hulette CM, et al. Microvascular injury and blood-brain barrier leakage in Alzheimer's disease. Neurobiology of aging 2007;28(7)977-86.

[58] Alafuzoff I, Seppo H, Mannermaa A, Riekkinen P Sr, Soininen H. Beta-amyloid load is not influenced by the severity of cardiovascular disease in aged and demented patients. Stroke 1999;30(3)613-8.

[59] Attems J, Jellinger KA, Lintner F. Alzheimer's disease pathology influences severity and topographical distribution of cerebral amyloid angiopathy. Acta neuropathologica 2005;110(3)222-31.

[60] Chalmers K, Wilcock GK, Love S. APOE epsilon 4 influences the pathological phenotype of Alzheimer's disease by favouring cerebrovascular over parenchymal accumulation of A beta protein. Neuropathology and applied neurobiology 2003;29(3)231-8.

[61] Christoforidis M, Schober R, Krohn K. Genetic-morphologic association study: Association between the low density lipoprotein-receptor related protein (LRP) and cerebral amyloid angiopathy. Neuropathology and applied neurobiology 2005;31(1)11-9.

[62] Chui HC, Zarow C, Mack WJ, Ellis WG, Zheng L, Jagust WJ, et al. Cognitive impact of subcortical vascular and Alzheimer's disease pathology. Annals of Neurology 2006;60(6)677-87.

[63] Davidson Y, Gibbons L, Pritchard A, Hardicre J, Wren J, Tian J, et al. Genetic associations between cathepsin D exon 2 C->T polymorphism and Alzheimer's disease, and pathological correlations with genotype. Journal of Neurology, Neurosurgery and Psychiatry 2006;77(4)515-7.

[64] Greenberg SM, Rebeck GW, Vonsattel JP, Gomez-Isla T, Hyman BT. Apolipoprotein E epsilon 4 and cerebral hemorrhage associated with amyloid angiopathy. Annals of Neurology 1995;38(2)254-9.

[65] Leclercq PD, Murray LS, Smith C, Graham DI, Nicoll JAR, Gentleman SM. Cerebral amyloid angiopathy in traumatic brain injury: Association with apolipoprotein E genotype. Journal of Neurology, Neurosurgery and Psychiatry 2005;76(2)229-33.

[66] Mortimer JA, Snowdon DA, Markesbery WR. The effect of APOE-e4 on dementia is mediated by Alzheimer neuropathology. Alzheimer Disease and Associated Disorders 2009;23(2)152-7.

[67] Tanskanen M, Lindsberg PJ, Tienari PJ, Polvikoski T, Sulkava R, Verkkoniemi A, et al. Cerebral amyloid angiopathy in a 95+cohort: Complement activation and apolipoprotein E (ApoE) genotype. Neuropathology and applied neurobiology 2005;31(6)589-99.

[68] Yip AG, Mckee AC, Green RC, Wells J, Young H, Cupples LA, et al. APOE, vascular pathology, and the AD brain. Neurology 2005;65(2)259-65.

[69] Zarow C, Zaias B, Lyness SA, Chui H. Cerebral amyloid angiopathy in Alzheimer disease is associated with apolipoprotein E4 and cortical neuron loss. Alzheimer Disease and Associated Disorders 1999;13(1)1-8.

[70] Zubenko GS, Stiffler S, Stabler S, Kopp U, Hughes HB, Cohen BM, et al. Association of the apolipoprotein E epsilon 4 allele with clinical subtypes of autopsy-confirmed Alzheimer's disease. American Journal of Medical Genetics 1994;54(3)199-205.

[71] Stroup DF, Berlin JA, Morton SC, Olkin I, Williamson GD, Rennie D, et al. Meta-analysis of observational studies in epidemiology: a proposal for reporting. Meta-analysis of Observational Studies in Epidemiology (MOOSE) group. JAMA: the journal of the American Medical Association 2000;283(15)2008-12.

[72] Little J, Higgins JPT, Ioannidis JPA, Moher D, Gagnon F, von Elm E, et al. STrenghtening the REporting of Genetic Association Studies (STREGA) – An Extension of the STROBE Statement. PLoS medicine 2009;6(2)e22.

[73] Rosenthal R. The file drawer problem and tolerance for null results. Psychological Bulletin 1979;86(3)638-41.

[74] Chen YW, Lee MJ, Smith EE. Cerebral amyloid angiopathy in East and West. International Journal of Stroke 2010;5(5)403-11.

[75] Attems J. Sporadic cerebral amyloid angiopathy: Pathology, clinical implications, and possible pathomechanisms. Acta neuropathologica 2005;110(4)345-59.

[76] Tzourio C, Arima H, Harrap S, Anderson C, Godin O, Woodward M, et al. APOE genotype, ethnicity, and the risk of cerebral hemorrhage. Neurology 2008;70(16)1322-8.

[77] Biffi A, Shulman JM, Jagiella JM, Cortellini L, Ayres AM, Schwab K, et al. Genetic variation at CR1 increases risk of cerebral amyloid angiopathy. Neurology 2012;78(5)334-41.

[78] Chalmers KA, Culpan D, Kehoe PG, Wilcock GK, Hughes A, Love S. APOE promoter, ACE1 and CYP46 polymorphisms and beta-amyloid in Alzheimer's disease. NeuroReport 2004;15(1)95-8.

[79] Chapuis J, Tian J, Shi J, Bensemain F, Cottel D, Lendon C, et al. Association study of the vascular endothelial growth factor gene with the risk of developing Alzheimer's disease. Neurobiology of aging 2006;27(9)1212-5.

[80] Chapuis J, Hot D, Hansmannel F, Kerdraon O, Ferreira S, Hubans C, et al. Transcriptomic and genetic studies identify IL-33 as a candidate gene for Alzheimer's disease. Molecular Psychiatry 2009;14(11)1004-16.

[81] Cuenco KT, Lunetta KL, Baldwin CT, Mckee AC, Guo J, Cupples LA, et al. Association of distinct variants in SORL1 with cerebrovascular and neurodegenerative changes related to Alzheimer disease. Archives of Neurology 2008;65(12)1640-8.

[82] Hamaguchi T, Okino S, Sodeyama N, Itoh Y, Takahashi A, Otomo E, et al. Association of a polymorphism of the transforming growth factor-beta1 gene with cerebral amyloid angiopathy. Journal of Neurology, Neurosurgery and Psychiatry 2005;76(5)696-9.

[83] Kölsch H, Larionov S, Dedeck O, Orantes M, Birkenmeier G, Griffin WST, et al. Association of the glutathione S-transferase omega-1 Ala140Asp polymorphism with cerebrovascular atherosclerosis and plaque-associated interleukin-1alpha expression. Stroke 2007;38(10)2847-50.

[84] Lendon CL, Thaker U, Harris JM, McDonagh AM, Lambert JC, Chartier-Harlin MC, et al. The angiotensin 1-converting enzyme insertion (I)/deletion (D) polymorphism does not influence the extent of amyloid or tau pathology in patients with sporadic Alzheimer's disease. Neuroscience Letters 2002;328(3)314-8.

[85] Peila R, Yucesoy B, White LR, Johnson V, Kashon ML, Wu K, et al. A TGF-beta1 polymorphism association with dementia and neuropathologies: the HAAS. Neurobiology of aging 2007;28(9)1367-73.

[86] Shi J, Tian J, Pritchard A, Lendon C, Lambert JC, Iwatsubo T, et al. A 3'-UTR polymorphism in the oxidized LDL receptor 1 gene increases Abeta40 load as cerebral amyloid angiopathy in Alzheimer's disease. Acta neuropathologica 2006;111(1)15-20.

[87] Thal DR, Papassotiropoulos A, Saido TC, Griffin WS, Mrak RE, Kölsch H, et al. Capillary cerebral amyloid angiopathy identifies a distinct APOE epsilon4-associated subtype of sporadic Alzheimer's disease. Acta neuropathologica 2010;120(2)169-83.

[88] Valant V, Keenan BT, Anderson CD, Shulman JM, Devan WJ, Ayres AM, et al. TOMM40 in cerebral amyloid angiopathy related intracerebral hemorrhage: comparative genetic analysis with Alzheimer's disease. Translational Stroke Research 2012;3(S1)S102-S112.

[89] Yamada M, Sodeyama N, Itoh Y, Suematsu N, Otomo E, Matsushita M, et al. Association of presenilin-1 polymorphism with cerebral amyloid angiopathy in the elderly. Stroke 1997;28(11)2219-21.

[90] Yamada M, Sodeyama N, Itoh Y, Suematsu N, Otomo E, Matsushita M, et al. Butyrylcholinesterase K variant and cerebral amyloid angiopathy. Stroke 1998;29(12)2488-90.

[91] Yamada M, Sodeyama N, Itoh Y, Suematsu N, Otomo E, Matsushita M, et al. Association of alpha1 antichymotrypsin polymorphism with cerebral amyloid angiopathy. Annals of Neurology 1998;44(1)129-31.

[92] Yamada M, Sodeyama N, Itoh Y, Suematsu N, Otomo E, Matsushita M, et al. A deletion polymorphism of alpha2 macroglobulin gene and cerebral amyloid angiopathy. Stroke 1999;30(11)2277-9.

[93] Yamada M, Sodeyama N, Itoh Y, Otomo E, Matsushita M, Mizusawa H. No association of paraoxonase genotype or atherosclerosis with cerebral amyloid angiopathy. Stroke 2002;33(4)896-900.

[94] Yamada M, Sodeyama N, Itoh Y, Takahashi A, Otomo E, Matsushita M, et al. Association of neprilysin polymorphism with cerebral amyloid angiopathy. Journal of Neurology Neurosurgery and Psychiatry 2003;74(6)749-51.

[95] Zubenko GS, Hughes HB, Stiffler JS. Clinical and neurobiological correlates of DXS1047 genotype in Alzheimer's disease. Biological Psychiatry 1999;46(2)173-81.

[96] Chibnik LB, Shulman JM, Leurgans SE, Schneider JA, Wilson RS, Tran D, et al. CR1 is associated with amyloid plaque burden and age-related cognitive decline. Annals of Neurology 2011;69(3)560-9.

[97] Brouwers HB, Biffi A, Ayres AM, Schwab K, Cortellini L, Romero JM, et al. Apolipoprotein E genotype predicts hematoma expansion in lobar intracerebral hemorrhage. Stroke 2012;43(6)1490-5.

[98] McCarron MO, Nicoll JAR. Apolipoprotein E genotype and cerebral amyloid angiopathy related hemorrhage. Annals of the New York Academy of Sciences 2000;903(176-9.

[99] Kalaria RN, Premkumar DRD. Apolipoprotein E genotype and cerebral amyloid angiopathy. Lancet 1995;346(8987)1424.

[100] Davis DG, Schmitt FA, Wekstein DR, Markesbery WR. Alzheimer neuropathologic alterations in aged cognitively normal subjects. Journal of Neuropathology and Experimental Neurology 1999;58(4)376-88.

[101] Verghese PB, Castellano JM, Holtzman DM. Apolipoprotein E in Alzheimer's disease and other neurological disorders. Lancet Neurology 2011;10(3)241-52.

[102] Dubois B, Feldman HH, Jacova C, Cummings JL, Dekosky ST, Barberger-Gateau P, et al. Revising the definition of Alzheimer's disease: a new lexicon. Lancet Neurology 2013;9(11)1118-27.

Chapter 6

Src Family Kinases in Intracerebral Hemorrhage

DaZhi Liu

Additional information is available at the end of the chapter

http://dx.doi.org/10.5772/58488

1. Introduction

Intracerebral hemorrhage (ICH), which accounts for 2 million (10–15%) of about 15 million strokes worldwide each year [1], has very high mortality rates of 31% at 7 days, 59% at 1 year, 82% at 10 years, and greater than 90% at 16 years [2,3]. ICH is associated with increased intracranial pressure, hematoma, blood brain barrier (BBB) disruption, brain edema, neuron loss, motor deficits, cognitive impairment and high mortality in humans. The major challenges immediately after ICH are re-bleed, hematoma induced brain injury, brain edema and neurological deficits [4]. Potential treatments of ICH include slowing the initial bleeding during the first hours after onset; removing blood from the parenchyma or ventricles to eliminate both mechanical and chemical factors that cause brain injury; management of complications of blood in the brain; and supportive medical care and surgery for certain patients [5]. Since these treatments have great variability, there is currently no FDA approved treatment for ICH.

The time course after ICH can be divided into two stages (acute/injury and chronic/recovery) (Fig. 1). At the acute stage, glutamine, thrombin, TNF-α, VEGF and other endogenous molecules are rapidly released following ICH. These molecules team up leading to brain cell death and severe brain injury via multiple neurotoxicity pathways, including (1) Excitatory amino acid (AA) and NMDA receptor-mediated excitatory toxicity; (2) Thrombin and other mitogen-mediated mitogenic stress; (3) Vascular endothelial growth factor (VEGF) and matrix metalloproteinases (MMPs)-mediated changes of vascular permeability; (4) Cytokines-mediated inflammatory responses; and others (Fig. 1).

As time course transits into chronic/recovery stage post-ICH, the elevated molecules resolve gradually and in turn participate in neurogenesis via populating neural progenitor cells (NPCs) to fix the damaged brain tissue. The possible mechanisms include (1) Excitatory amino acid (AA) and NMDA receptor-mediated excitatory genesis; (2) Thrombin and other mitogen-mediated mitogenic growth; (3) Nerve growth factor (NGF), epidermal growth factor (EGF) and other growth factor-mediated neurogenesis; and others (Fig. 1).

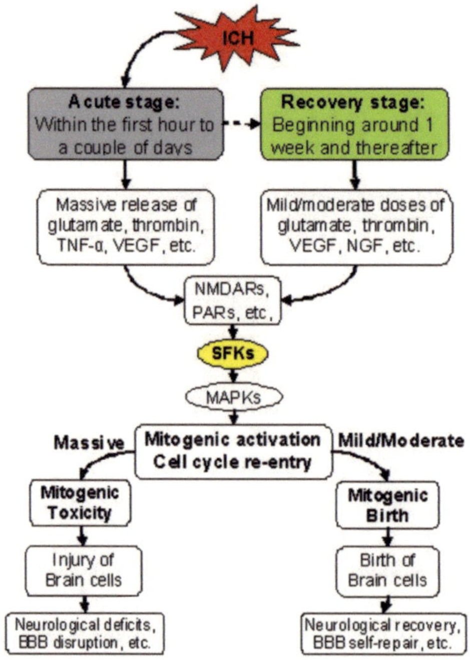

Figure 1. Src family kinases (SFKs) participate in mitogenic signaling pathways that play critical roles in blood-brain barrier (BBB) disruption and self-repair after intracerebral hemorrhage (ICH). The network does not apply to certain cell type(s). The arrows do not necessarily indicate direct binding and/or activation of the downstream molecules; intermediate proteins or kinases may exist.

Src family kinases (SFKs), a family of proto-oncogenic proteins, participate in both neurotoxicity at acute stage and neurogenesis during recovery stage post-ICH (Fig 1). Our previous studies have confirmed the time-specific and conflicting roles of SFKs using their inhibitor (PP2) after ICH: (1) Acute administration of PP2 (immediately post-ICH) decreases local cerebral glucose utilization (LCGU), activity of SFKs, attenuates BBB breakdown, brain edema, and cell death around ICH and improves behavioral function following ICH [6-9]; (2) Chronic inhibition of PP2 (2-6 days) blocks BBB repair and brain edema resolution in the recovery stage (7-14 days) after ICH [9].

2. Tissue specificity, structure and activity regulation of SFKs

SFKs are a family of non-receptor protein tyrosine kinases, include nine family members Src, Fyn, Lck, Lyn, Yes, Hck, Blk, Fgr, and Yrk [10-12], of which Src, Fyn, Yes and Yrk widely expressed whereas the rest members are expressed in specific tissues [13]. In addition, one tissue can express multiple SFK members, for example, Src, Fyn, Yes, and Lck have been examined in brain [13-19]. Importantly, the different SFK family members often compensate for one another [20], which are supported by the evidence that the mice deficient in Src can survive though Src plays vital role in cell signaling transduction [20]. SFKs share a conserved domain structure consisting of consecutive SH3 (polyproline type II helix for protein-protein interaction), SH2 (phosphotyrosine recognition), and SH1 (tyrosine kinase catalytic activity) [12]. All SFK family members also contain an membrane-targeting region at their N-terminus that is followed by a unique domain of 50–70 residues, and the unique region is divergent among family members [12]. Although it still remains incompletely clear, Src activity is regulated by tyrosine phosphorylation at two sites (one is at Tyr416 in the SH1 domain, the other at Tyr527 in the short C-terminal tail), but with opposing effects. While phosphorylation at Tyr416 activates Src, phosphorylation at Tyr527 inactivates Src [13,21].

3. SFKs modulate NMDA receptor for brain injury after ICH

NMDA receptors are ionotropic glutamate receptors, comprise NR1, NR2 and NR3 subunits, which form the central conductance pathway [22,23]. In the physiological conductions, activation of NMDA receptors results in the opening of an ion channel that allows the flow of Na^+ and small amounts of Ca^{2+} into the cell and K^+ out of the cell [22,23]. Following ICH there is a transient increase of glutamate release and local cerebral glucose utilization in the region surrounding the ICH, and the antagonists of NMDA receptors reverse the glucose hypermetabolism produced by ICH [6]. However, glutamate alone could not explain the hypermetabolism since glutamate injected directly into brain did not produce hypermetabolism [6]. Apart from glutamate release, ICH may affect NMDA receptors in some way to make them more sensitive to glutamate in order to mediate injury and/or hypermetabolism.

ICH activates SFKs [7,8], and SFK members (e.g., Src, Fyn) up-regulate the ion channel activity of NMDA receptor and make them more sensitive to glutamate by phosphorylating the NR2A and NR2B subunits of the NMDA receptors [19,24-27]. It has been proved that phosphorylation by Src at Tyr-1292, Tyr-1325 and Tyr-1387 in NR2A subunit increases activity of NMDA receptors, and phosphorylation of tyrosine residues by Src in the C-terminal of the subunits prevents a Zn^{2+}-dependent inhibition of the NMDA receptors and thus increases channel conductivity [28-30]. We have demonstrated that either NMDA receptor inhibitors (MK-801) or SKF inhibitors (PP2) can attenuate brain injury at the acute stage after ICH (Fig. 2) [8]. These results suggest that either activation of NMDA receptors or SFKs is sufficient to produce brain injury post-ICH.

NMDA receptor activation has also been shown to enhance NPCs proliferation and lead to increased neurogenesis [31]. However, there is no direct report showing the mechanism by which SFKs participate in NMDA receptors mediated neurogenesis after brain injury.

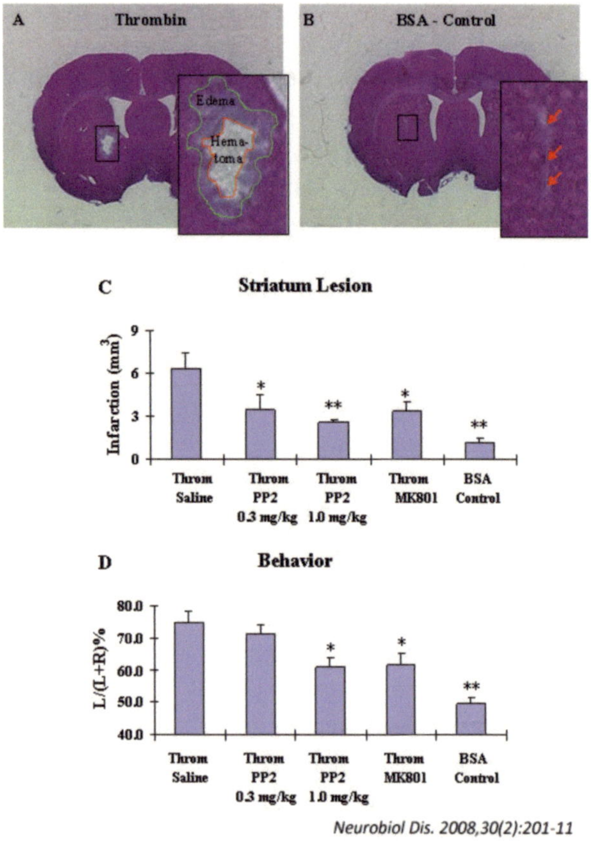

Figure 2. The effects of graded doses of SFK inhibitor PP2 (0.3 and 1.0 mg/kg, i.p.) and NMDA receptor inhibitor MK801 (1.0 mg/kg, i.p.) on injury produced by thrombin injections into striatum of Sprague-Dawley rats. **A.** Representative Hematoxylin-Eosin stained section shows that 20U of thrombin causes brain injury, including hematoma and edema. **B.** Control injections of BSA into striatum produced minor injury. **C.** Infarction volumes 24h following striatum injections of thrombin compared to control BSA injections (BSA/Control) (n=6). Several groups of animals received striatal injections of thrombin: just thrombin alone (Throm/Saline) (n=6); prior intraperitoneal injection of PP2 (Throm/PP2/0.3mg/kg) (n=6); prior intraperitoneal injection of a higher dose of PP2 (Throm/PP2/1.0mg/kg) (n=6); and prior intraperitoneal injection of MK801 (Throm/MK801/1.0mg/kg) (n=6).. * $p<0.05$ and ** $P<0.01$ vs Throm/Saline (one-way ANOVA followed by Dunnett's *post hoc* test). **D.** PP2 (0.3 and 1.0 mg/kg, i.p.) and MK801 (1.0 mg/kg, i.p.) decrease the thrombin-induced motor deficits (n=9 for each group) using Elevated Body Swing Test (EBST) 23.5 hours after thrombin injections. Biased swinging behavior was calculated as follows: L/ (L+R) (%) for left biased swings (L, left-biased swings; R, right-biased swings). * $p<0.05$ and ** $P<0.01$ vs Throm/Saline (one-way ANOVA followed by Dunnett's *post hoc* test).

4. SFKs regulate cell cycle for mitogenic toxicity and cell proliferation after ICH

The cell cycle is an irreversible, ordered set of events that normally leads to cellular division [32-36]. The release of cells from a quiescent state (G0) results in their entry into the first gap phase (G1), during which the cells prepare for DNA replication in the synthetic phase (S). This is followed by the second gap phase (G2) and mitosis phase (M). After the cell has split into its two daughter cells, the new cells enter either G1 or G0. Mature neurons normally maintain themselves in G0 resting phase; however, a mature neuron that re-enters the cell cycle can neither advance to a new G0 quiescent state nor revert to its earlier G0 state. This presents a critical dilemma to the neuron from which death may be an unavoidable, but necessary, outcome for adult neurons attempting to complete the cell cycle [32,37]. Increasing evidence have revealed that aberrant cell cycle re-entry leads to neuronal death [8,32,37-64], and cell cycle inhibition via blocking SFKs can protect neurons from death post-ICH [8].

Apart from post-mitotic neurons, SFKs play critical roles in the process of cell cycle in dividable cells, by regulating mitogen-activated protein kinases (MAPKs) and cell cycle proteins such as cyclin-dependent kinases (Cdks) [65-69]. Although mitogenic signaling is necessary to initiate the cell cycle for normal cell division and proliferation, massive mitogenic signaling can also produce neurotoxicity and cell death [9,32,37,70]. Cell death and cell proliferation seem contradictory to each other, but these two seemingly different cellular processes share some common mitogenic molecules and signaling pathways (Fig. 1). In addition, many other molecules, including Ca^{2+}, ROS, NO and MMPs can directly or indirectly activate or increase mitogenic signaling [54,71-77].

There are two stages (acute and recovery) after ICH (Fig. 1). Once ICH occurs, a large number of molecules (e.g., thrombin, glutamate, TNF-α, VEGF, etc.) are increased. This peaks within the first hour to a day in the acute stage after ICH, and then resolves gradually in the recovery stage after ICH. The over-activated SFKs/mitogenic signaling leads to neurons to enter the cell cycle and die, and damages astroctyes and BMVECs via MAPKs in acute stage after ICH. Within a day, however, the massive thrombin/SFK mitogenic signaling resolve, and the disease progresses to a recovery stage of ICH. The restored moderate SFK/mitogenic signaling leads to newborn BMVECs, astrocytes and other cells that mediate self-repair in the recovery stage after ICH.

As shown in Fig. 1, thrombin (a potent mitogen) triggers mitosis after ICH by modulating mitogenic intracellular molecules such as SFKs. SFKs participate in mitogenic signaling activation via regulating mitogen-activated protein kinases (MAPKs) and other molecules [64-69] that play critical roles not only in brain injuries during the acute stage in ICH, but in brain self-repair during the recovery stage in ICH. Acute inhibition of SFKs is beneficial, that attenuates hematoma, BBB breakdown, vasogenic edema, MAPK activation in the acute stage (0-24h) after ICH (Fig. 2, 3, 4, 5 & 6) [7-9,26,64,70]. In contrast, delayed and lasting inhibition of SFKs is detrimental, and prolongs BBB repair and brain edema resolution in the recovery stage (7-14 days) after ICH [9], presumably because SFKs mediate population of NPCs that exist in the "neurovascular niche". that repair the damaged BBB [78]. Such NPCs could serve as a source of newborn cells (i.e., BMVECs, astrocytes and perhaps other cells) of the neurovascular unit that play a role in re-establishing the BBB via the mitogenic growth signaling pathways during recovery phase after ICH (Fig. 1) [79].

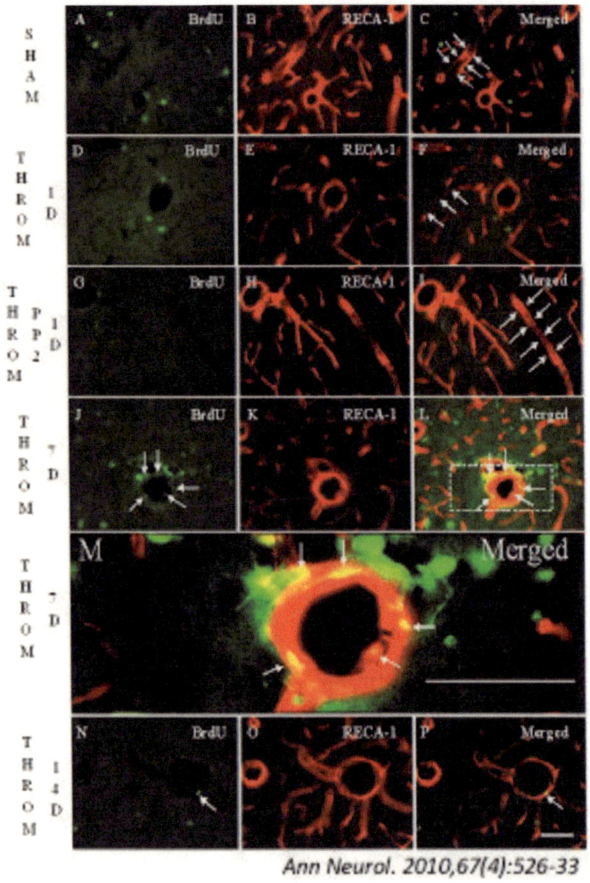

Figure 3. Intracerebroventricular (i.c.v.) injection of thrombin (20 U/animal, i.c.v.) causes reductions in brain microvascular endothelial cell (BMVEC) immunoreactivity after 1 day, and subsequent BMVEC proliferation around the brain vessels in lacunosum moleculare layer (LMol) of the hippocampus after 7 days and 14 days in Sprague-Dawley rats. **Panels A-C** show rats following sham operations labeled for BrdU, bromodeoxyuridine (A), RECA-1, rat endothelial antigen-1 (B) and the overlay or Merged image (C). RECA-1$^+$cells demonstrate the tube-shape of brain capillaries (arrows in panel C). **Panels D to F** show BrdU (D), RECA-1 (E) and the Merged image (F) at 1 day after thrombin injections. Compared with the sham group, RECA-1$^+$cells tend to lose their tube-shape at 1 day following thrombin injections (arrows in panel F) and there were no BrdU$^+$cells co-labeled with RECA-1 at one day (panel F). **Panels G-I** show the staining for BrdU (G), RECA-1 (H) and the Merged image (I) 1 day after thrombin plus PP2 injections. PP2 administration at day 0, immediately after thrombin injection, blocks the thrombin-induced loss of tube-shape of RECA-1$^+$cells. **Panels J-L** show the staining of BrdU (J), RECA-1 (K) and the Merged image (L) 7 days after thrombin injection. Compared to 1 day, BrdU$^+$cells are increased 7 days after thrombin injection. Some of these BrdU$^+$cells are co-labeled with RECA-1 (arrows in panel L and M). A few brain capillaries regained their tube-shape, though not completely. **Panel M** shows a higher power image of Panel L (area within dashed lines). RECA-1 stained BMVEC are red. The BrdU$^+$/RECA-1$^+$double-labeled new born BMVEC nuclei are yellow. **Panels N to P** show the staining for BrdU (M), RECA-1 (N) and the Merged image (P) 14 days after thrombin injection. Compared to 7 days, BrdU$^+$cells are decreased, but some BrdU$^+$cells (N) remain co-labeled with RECA-1 (arrow in panel P), and more and more brain capillaries regained the tube-shape 14 days after the thrombin injection. Scale bars: A-P, 50 μm.

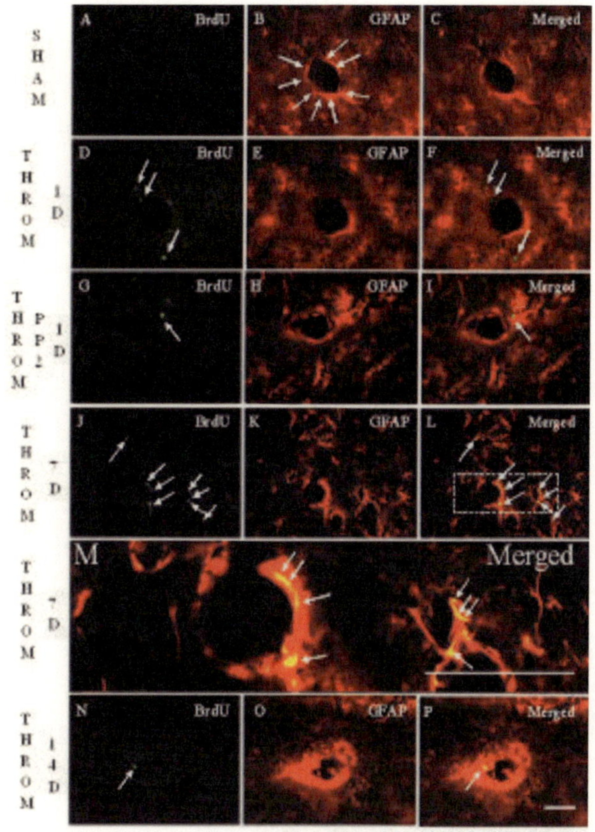

Ann Neurol. 2010,67(4):526-33

Figure 4. Intracerebroventricular injection of thrombin (20 U/animal, i.c.v.) causes reductions in astrocyte glial fibrillary acidic protein (GFAP) immunoreactivity after 1 day, and subsequent astrocyte proliferation around the brain vessels in lacunosum moleculare layer (LMol) of the hippocampus after 7 days and 14 days in Sprague-Dawley rats. **Panels A-C** show rats with sham operation labeled for BrdU (A), GFAP (B), and Merged image (C). GFAP$^+$cells envelop most all of the brain vessel (arrows in panel B). **Panels D-F** show BrdU (D), GFAP (E) and the Merged image (F) at 1 day after thrombin injection. Compared with the sham group, there is decreased GFAP immunoreactivity around brain vessels. There are a few BrdU$^+$/GFAP$^-$cells located close to the vessel (arrows in panel F). **Panels G-I** show the staining for BrdU (G), GFAP (H) and the Merged image (I) 1 day after thrombin plus PP2 injections. PP2 administration at day 0, immediately after thrombin injection, blocks the thrombin-induced reductions in GFAP immunoreactivity. **Panels J-L** show the staining for BrdU (J), GFAP (K) and the Merged image (L) 7 days after thrombin injection. Compared to 1 day, BrdU $^+$cells are increased 7 days after thrombin injection (J, arrows). Many of these BrdU$^+$cells are co-labeled with GFAP (arrows in panel L). **Panel M** shows a higher power image of Panel L (area within dashed lines). GFAP stained astrocytes are red. The BrdU$^+$GFAP$^+$double-labeled new born astrocytic nuclei are yellow (arrows, Panel M). **Panels N-P** show the staining for BrdU (N), GFAP (O) and the Merged image (P) 14 days after thrombin injection. Compared to 7 days, BrdU $^+$cells are decreased 14 days after thrombin injection. Some BrdU$^+$cells (N, arrow) remain co-labeled with GFAP (arrow in panel P) 14 days after the thrombin injection. Scale bars: A-P, 50 μm.

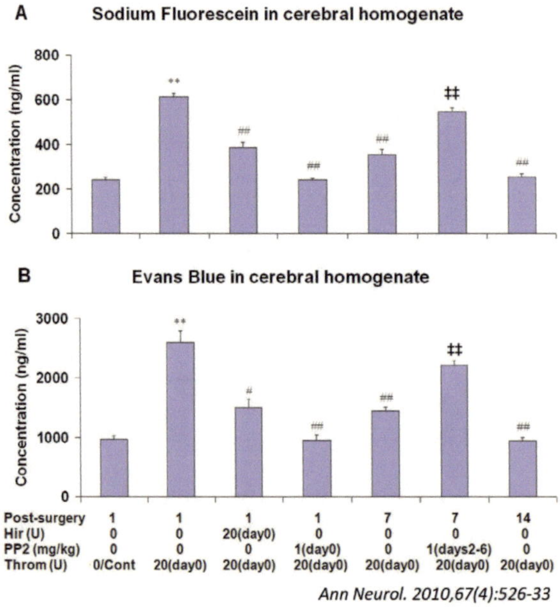

Figure 5. Brain sodium fluorescein (NF, **panel A**) and Evans blue (EB, **panel B**) extravasation increased 1 day after thrombin (Throm) injections (20U, i.c.v.), and decreased at 7 and 14 days in Sprague-Dawley rats. The thrombin inhibitor hirudin (Hir, 20U) blocked thrombin-induced NF/EB extravasation at 1 day after co-injection into the cerebral ventricles. PP2 (src family kinase inhibitor) administered with thrombin (day 0) blocked the NF/EB extravasation at 1 day after thrombin injection, whereas delayed PP2 administration (days 2-6) postponed alleviation of NF/EB extravasation at 7 days post-thrombin injection. Each column and vertical bar represents the mean ± standard error of the mean. ** $p<0.01$ vs. Cont; #$p<0.05$, ##$p<0.01$ vs. Throm/1day, ‡‡ $p<0.01$ vs. Throm/7days (one-way ANOVA followed by Tukey's *post hoc* test).

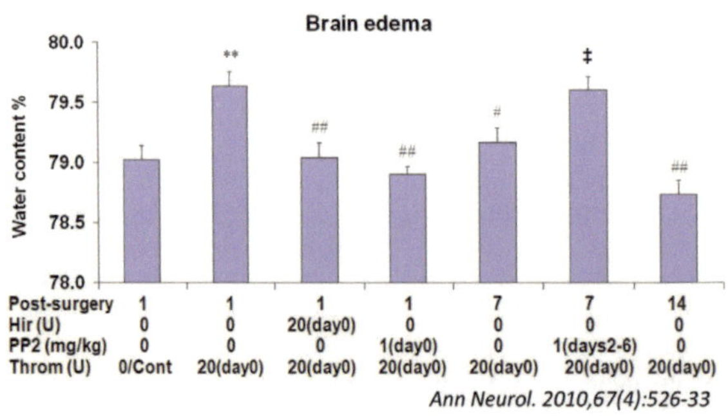

Figure 6. Brain edema (water content) increased at 1 day after thrombin (Throm) injections (20U, i.c.v.), and decreased by 7 and 14 days in Sprague-Dawley rats. The thrombin inhibitor hirudin (Hir, 20U, i.c.v.) blocked elevation thrombin-induced brain water content at 1 day after co-injection into the cerebral ventricle. Administration of PP2 (src family kinase inhibitor) at day 0 blocked the increase in brain water content observed at 1 day after thrombin injection, whereas delayed PP2 administration (days 2-6) prevented the resolution of brain water content at 7 days post-thrombin injection. Each column and vertical bar represents the mean ± standard error of the mean. ** $p<0.01$ vs. Cont; #$p<0.05$, ##$p<0.01$ vs. Throm/1day, ‡ $p<0.05$ vs. Throm/7days (one-way ANOVA followed by Tukey's *post hoc* test).

5. Future directions

Future studies need to address which specific SFK members found in brain (e.g., Src, Fyn, Lck and Yrk) that mediate ICH-induced cell death or birth. Since delayed and chronic inhibition of SFKs may impair neurogenesis and prolong BBB self-repair during recovery stage post-ICH, the acute and transient inhibition of SFKs should be pursued in treatment of ICH. The nanoparticle-based siRNA transfection system allows transient knockdown of target gene(s) and highly efficient delivery of siRNA *in vivo* with low cytotoxicity [80,81]. This could present a novel therapy for treating ICH patients as the nanoparticle-based siRNA approach provides heightened specificity for specific SFK gene(s) with less off target effects and this approach has been used in humans [82-85].

Acknowledgements

The authors acknowledge the support of AHA Beginning Grant-in-Aid 12BGIA12060381 (DZL) and NIH grant NS054652 (FRS). Figures 2, and 3-6 were published in *Neurobiol Dis. 2008,30(2):201-11* and *Ann Neurol.* 2010,67(4):526-33, respectively.

Author details

DaZhi Liu*

Address all correspondence to: dzliu@ucdavis.edu

Department of Neurology and the M.I.N.D. Institute, University of California at Davis Medical Center, Sacramento, California, USA

References

[1] Sudlow CL, Warlow CP (1997) Comparable studies of the incidence of stroke and its pathological types: results from an international collaboration. International Stroke Incidence Collaboration. Stroke 28: 491-499.

[2] Flaherty ML, Haverbusch M, Sekar P, Kissela B, Kleindorfer D, et al. (2006) Long-term mortality after intracerebral hemorrhage. Neurology 66: 1182-1186.

[3] Fogelholm R, Murros K, Rissanen A, Avikainen S (2005) Long term survival after primary intracerebral haemorrhage: a retrospective population based study. J Neurol Neurosurg Psychiatry 76: 1534-1538.

[4] Gebel JM, Jr., Jauch EC, Brott TG, Khoury J, Sauerbeck L, et al. (2002) Natural history of perihematomal edema in patients with hyperacute spontaneous intracerebral hemorrhage. Stroke 33: 2631-2635.

[5] Broderick J, Connolly S, Feldmann E, Hanley D, Kase C, et al. (2007) Guidelines for the management of spontaneous intracerebral hemorrhage in adults: 2007 update: a guideline from the American Heart Association/American Stroke Association Stroke Council, High Blood Pressure Research Council, and the Quality of Care and Outcomes in Research Interdisciplinary Working Group. Circulation 116: e391-413.

[6] Ardizzone TD, Lu A, Wagner KR, Tang Y, Ran R, et al. (2004) Glutamate receptor blockade attenuates glucose hypermetabolism in perihematomal brain after experimental intracerebral hemorrhage in rat. Stroke 35: 2587-2591

[7] Ardizzone TD, Zhan X, Ander BP, Sharp FR (2007) SRC kinase inhibition improves acute outcomes after experimental intracerebral hemorrhage. Stroke 38: 1621-1625.

[8] Liu DZ, Cheng XY, Ander BP, Xu H, Davis RR, et al. (2008) Src kinase inhibition decreases thrombin-induced injury and cell cycle re-entry in striatal neurons. Neurobiol Dis 30: 201-211.

[9] Liu DZ, Ander BP, Xu H, Shen Y, Kaur P, et al. (2010) Blood-brain barrier breakdown and repair by Src after thrombin-induced injury. Ann Neurol 67: 526-533.

[10] Oda H, Kumar S, Howley PM (1999) Regulation of the Src family tyrosine kinase Blk through E6AP-mediated ubiquitination. Proc Natl Acad Sci U S A 96: 9557-9562.

[11] Biscardi JS, Ishizawar RC, Silva CM, Parsons SJ (2000) Tyrosine kinase signalling in breast cancer: epidermal growth factor receptor and c-Src interactions in breast cancer. Breast Cancer Res 2: 203-210.

[12] Boggon TJ, Eck MJ (2004) Structure and regulation of Src family kinases. Oncogene 23: 7918-7927.

[13] Parsons SJ, Parsons JT (2004) Src family kinases, key regulators of signal transduction. Oncogene 23: 7906-7909.

[14] Morse WR, Whitesides JG, 3rd, LaMantia AS, Maness PF (1998) p59fyn and pp60c-src modulate axonal guidance in the developing mouse olfactory pathway. J Neurobiol 36: 53-63.

[15] Encinas M, Tansey MG, Tsui-Pierchala BA, Comella JX, Milbrandt J, et al. (2001) c-Src is required for glial cell line-derived neurotrophic factor (GDNF) family ligand-mediated neuronal survival via a phosphatidylinositol-3 kinase (PI-3K)-dependent pathway. J Neurosci 21: 1464-1472.

[16] Sperber BR, Boyle-Walsh EA, Engleka MJ, Gadue P, Peterson AC, et al. (2001) A unique role for Fyn in CNS myelination. J Neurosci 21: 2039-2047.

[17] Heidinger V, Manzerra P, Wang XQ, Strasser U, Yu SP, et al. (2002) Metabotropic glutamate receptor 1-induced upregulation of NMDA receptor current: mediation

through the Pyk2/Src-family kinase pathway in cortical neurons. J Neurosci 22: 5452-5461.

[18] Rouer E (2010) [Neuronal isoforms of Src, Fyn and Lck tyrosine kinases: A specific role for p56lckN in neuron protection]. C R Biol 333: 1-10.

[19] Salter MW, Kalia LV (2004) Src kinases: a hub for NMDA receptor regulation. Nat Rev Neurosci 5: 317-328.

[20] Stein PL, Vogel H, Soriano P (1994) Combined deficiencies of Src, Fyn, and Yes tyrosine kinases in mutant mice. Genes Dev 8: 1999-2007.

[21] Hunter T (1987) A tail of two src's: mutatis mutandis. Cell 49: 1-4.

[22] McBain CJ, Mayer ML (1994) N-methyl-D-aspartic acid receptor structure and function. Physiol Rev 74: 723-760.

[23] Frank RA (2011) Endogenous ion channel complexes: the NMDA receptor. Biochem Soc Trans 39: 707-718.

[24] Ali DW, Salter MW (2001) NMDA receptor regulation by Src kinase signalling in excitatory synaptic transmission and plasticity. Curr Opin Neurobiol 11: 336-342.

[25] Groveman BR, Feng S, Fang XQ, Pflueger M, Lin SX, et al. (2012) The regulation of N-methyl-D-aspartate receptors by Src kinase. FEBS J 279: 20-28.

[26] Sharp F, Liu DZ, Zhan X, Ander BP (2008) Intracerebral hemorrhage injury mechanisms: glutamate neurotoxicity, thrombin, and Src. Acta Neurochir Suppl 105: 43-46.

[27] Hossain MI, Kamaruddin MA, Cheng HC (2012) Aberrant regulation and function of Src family tyrosine kinases: their potential contributions to glutamate-induced neurotoxicity. Clin Exp Pharmacol Physiol 39: 684-691.

[28] Yang M, Leonard JP (2001) Identification of mouse NMDA receptor subunit NR2A C-terminal tyrosine sites phosphorylated by coexpression with v-Src. J Neurochem 77: 580-588.

[29] Zheng F, Gingrich MB, Traynelis SF, Conn PJ (1998) Tyrosine kinase potentiates NMDA receptor currents by reducing tonic zinc inhibition. Nat Neurosci 1: 185-191.

[30] Kohr G, Seeburg PH (1996) Subtype-specific regulation of recombinant NMDA receptor-channels by protein tyrosine kinases of the src family. J Physiol 492 (Pt 2): 445-452.

[31] Suzuki M, Nelson AD, Eickstaedt JB, Wallace K, Wright LS, et al. (2006) Glutamate enhances proliferation and neurogenesis in human neural progenitor cell cultures derived from the fetal cortex. Eur J Neurosci 24: 645-653.

[32] Liu DZ, Ander BP, Sharp FR (2010) Cell cycle inhibition without disruption of neurogenesis is a strategy for treatment of central nervous system diseases. Neurobiol Dis 37: 549-557.

[33] Novak B, Tyson JJ, Gyorffy B, Csikasz-Nagy A (2007) Irreversible cell-cycle transitions are due to systems-level feedback. Nat Cell Biol 9: 724-728.

[34] Norbury C, Nurse P (1992) Animal cell cycles and their control. Annu Rev Biochem 61: 441-470.

[35] Malumbres M, Barbacid M (2001) To cycle or not to cycle: a critical decision in cancer. Nat Rev Cancer 1: 222-231.

[36] Schwartz GK, Shah MA (2005) Targeting the cell cycle: a new approach to cancer therapy. J Clin Oncol 23: 9408-9421.

[37] Liu DZ, Ander BP (2012) Cell cycle inhibition without disruption of neurogenesis is a strategy for treatment of aberrant cell cycle diseases: an update. ScientificWorldJournal 2012: 491737.

[38] Nagy Z, Esiri MM, Cato AM, Smith AD (1997) Cell cycle markers in the hippocampus in Alzheimer's disease. Acta Neuropathol 94: 6-15.

[39] McShea A, Harris PL, Webster KR, Wahl AF, Smith MA (1997) Abnormal expression of the cell cycle regulators P16 and CDK4 in Alzheimer's disease. Am J Pathol 150: 1933-1939.

[40] Nagy Z, Esiri MM (1998) Neuronal cyclin expression in the hippocampus in temporal lobe epilepsy. Exp Neurol 150: 240-247.

[41] Busser J, Geldmacher DS, Herrup K (1998) Ectopic cell cycle proteins predict the sites of neuronal cell death in Alzheimer's disease brain. J Neurosci 18: 2801-2807.

[42] Copani A, Condorelli F, Caruso A, Vancheri C, Sala A, et al. (1999) Mitotic signaling by beta-amyloid causes neuronal death. Faseb J 13: 2225-2234.

[43] Yang Y, Geldmacher DS, Herrup K (2001) DNA replication precedes neuronal cell death in Alzheimer's disease. J Neurosci 21: 2661-2668.

[44] Copani A, Uberti D, Sortino MA, Bruno V, Nicoletti F, et al. (2001) Activation of cell-cycle-associated proteins in neuronal death: a mandatory or dispensable path? Trends Neurosci 24: 25-31.

[45] O'Hare M, Wang F, Park DS (2002) Cyclin-dependent kinases as potential targets to improve stroke outcome. Pharmacol Ther 93: 135-143.

[46] Lefevre K, Clarke PG, Danthe EE, Castagne V (2002) Involvement of cyclin-dependent kinases in axotomy-induced retinal ganglion cell death. J Comp Neurol 447: 72-81.

[47] Imai H, Harland J, McCulloch J, Graham DI, Brown SM, et al. (2002) Specific expression of the cell cycle regulation proteins, GADD34 and PCNA, in the peri-infarct zone after focal cerebral ischaemia in the rat. Eur J Neurosci 15: 1929-1936.

[48] Jordan-Sciutto KL, Dorsey R, Chalovich EM, Hammond RR, Achim CL (2003) Expression patterns of retinoblastoma protein in Parkinson disease. J Neuropathol Exp Neurol 62: 68-74.

[49] Yang Y, Mufson EJ, Herrup K (2003) Neuronal cell death is preceded by cell cycle events at all stages of Alzheimer's disease. J Neurosci 23: 2557-2563.

[50] Ranganathan S, Bowser R (2003) Alterations in G(1) to S phase cell-cycle regulators during amyotrophic lateral sclerosis. Am J Pathol 162: 823-835.

[51] Nguyen MD, Boudreau M, Kriz J, Couillard-Despres S, Kaplan DR, et al. (2003) Cell cycle regulators in the neuronal death pathway of amyotrophic lateral sclerosis caused by mutant superoxide dismutase 1. J Neurosci 23: 2131-2140.

[52] Herrup K, Neve R, Ackerman SL, Copani A (2004) Divide and die: cell cycle events as triggers of nerve cell death. J Neurosci 24: 9232-9239.

[53] Kuan CY, Schloemer AJ, Lu A, Burns KA, Weng WL, et al. (2004) Hypoxia-ischemia induces DNA synthesis without cell proliferation in dying neurons in adult rodent brain. J Neurosci 24: 10763-10772.

[54] Copani A, Nicoletti F (2005) Cell-cycle mechanisms and neuronal cell death. New York: Kluwer Academic/Plenum.

[55] Di Giovanni S, Movsesyan V, Ahmed F, Cernak I, Schinelli S, et al. (2005) Cell cycle inhibition provides neuroprotection and reduces glial proliferation and scar formation after traumatic brain injury. Proc Natl Acad Sci U S A 102: 8333-8338.

[56] Andorfer C, Acker CM, Kress Y, Hof PR, Duff K, et al. (2005) Cell-cycle reentry and cell death in transgenic mice expressing nonmutant human tau isoforms. J Neurosci 25: 5446-5454.

[57] Copani A, Hoozemans JJ, Caraci F, Calafiore M, Van Haastert ES, et al. (2006) DNA polymerase-beta is expressed early in neurons of Alzheimer's disease brain and is loaded into DNA replication forks in neurons challenged with beta-amyloid. J Neurosci 26: 10949-10957.

[58] Zhang Y, Qu D, Morris EJ, O'Hare MJ, Callaghan SM, et al. (2006) The Chk1/Cdc25A pathway as activators of the cell cycle in neuronal death induced by camptothecin. J Neurosci 26: 8819-8828.

[59] Yang Y, Varvel NH, Lamb BT, Herrup K (2006) Ectopic cell cycle events link human Alzheimer's disease and amyloid precursor protein transgenic mouse models. J Neurosci 26: 775-784.

[60] Khurana V, Lu Y, Steinhilb ML, Oldham S, Shulman JM, et al. (2006) TOR-mediated cell-cycle activation causes neurodegeneration in a Drosophila tauopathy model. Curr Biol 16: 230-241.

[61] Rao HV, Thirumangalakudi L, Desmond P, Grammas P (2007) Cyclin D1, cdk4, and Bim are involved in thrombin-induced apoptosis in cultured cortical neurons. J Neurochem 101: 498-505.

[62] Zhang J, Herrup K (2011) Nucleocytoplasmic Cdk5 is involved in neuronal cell cycle and death in post-mitotic neurons. Cell Cycle 10: 1208-1214.

[63] Liu DZ, Ander BP (2011) Cell cycle phase transitions: signposts for aberrant cell cycle reentry in dying mature neurons. Journal of Cytology & Histology 2: 5.

[64] (2004) Finishing the euchromatic sequence of the human genome. Nature 431: 931-945.

[65] Grimmler M, Wang Y, Mund T, Cilensek Z, Keidel EM, et al. (2007) Cdk-inhibitory activity and stability of p27Kip1 are directly regulated by oncogenic tyrosine kinases. Cell 128: 269-280.

[66] Kasahara K, Nakayama Y, Nakazato Y, Ikeda K, Kuga T, et al. (2007) Src signaling regulates completion of abscission in cytokinesis through ERK/MAPK activation at the midbody. J Biol Chem 282: 5327-5339.

[67] Liu Z, Falola J, Zhu X, Gu Y, Kim LT, et al. (2004) Antiproliferative effects of Src inhibition on medullary thyroid cancer. J Clin Endocrinol Metab 89: 3503-3509.

[68] Mishra R, Wang Y, Simonson MS (2005) Cell cycle signaling by endothelin-1 requires Src nonreceptor protein tyrosine kinase. Mol Pharmacol 67: 2049-2056.

[69] Taylor SJ, Shalloway D (1993) The cell cycle and c-Src. Curr Opin Genet Dev 3: 26-34.

[70] Liu DZ, Sharp FR (2011) The dual role of SRC kinases in intracerebral hemorrhage. Acta Neurochir Suppl 111: 77-81.

[71] Li N, Wang C, Wu Y, Liu X, Cao X (2009) Ca2+/Calmodulin-dependent Protein Kinase II Promotes Cell Cycle Progression by Directly Activating MEK1 and Subsequently Modulating p27 Phosphorylation. J Biol Chem 284: 3021-3027.

[72] Mao LM, Tang QS, Wang JQ (2009) Regulation of extracellular signal-regulated kinase phosphorylation in cultured rat striatal neurons. Brain Res Bull 78: 328-334.

[73] de Bernardo S, Canals S, Casarejos MJ, Solano RM, Menendez J, et al. (2004) Role of extracellular signal-regulated protein kinase in neuronal cell death induced by glutathione depletion in neuron/glia mesencephalic cultures. J Neurochem 91: 667-682.

[74] Qian JY, Leung A, Harding P, LaPointe MC (2006) PGE2 stimulates human brain natriuretic peptide expression via EP4 and p42/44 MAPK. Am J Physiol Heart Circ Physiol 290: H1740-1746.

[75] Fiebich BL, Schleicher S, Spleiss O, Czygan M, Hull M (2001) Mechanisms of prostaglandin E2-induced interleukin-6 release in astrocytes: possible involvement of EP4-

like receptors, p38 mitogen-activated protein kinase and protein kinase C. J Neurochem 79: 950-958.

[76] Meissirel C, Ruiz de Almodovar C, Knevels E, Coulon C, Chounlamountri N, et al. (2011) VEGF modulates NMDA receptors activity in cerebellar granule cells through Src-family kinases before synapse formation. Proc Natl Acad Sci U S A 108: 13782-13787.

[77] Rosenberg GA (2009) Matrix metalloproteinases and their multiple roles in neurodegenerative diseases. Lancet Neurol 8: 205-216.

[78] Ohab JJ, Fleming S, Blesch A, Carmichael ST (2006) A neurovascular niche for neurogenesis after stroke. J Neurosci 26: 13007-13016.

[79] Palmer TD, Willhoite AR, Gage FH (2000) Vascular niche for adult hippocampal neurogenesis. J Comp Neurol 425: 479-494.

[80] Hannon GJ (2002) RNA interference. Nature 418: 244-251.

[81] Reynolds A, Leake D, Boese Q, Scaringe S, Marshall WS, et al. (2004) Rational siRNA design for RNA interference. Nat Biotechnol 22: 326-330.

[82] de Fougerolles A, Vornlocher HP, Maraganore J, Lieberman J (2007) Interfering with disease: a progress report on siRNA-based therapeutics. Nat Rev Drug Discov 6: 443-453.

[83] Eifler AC, Thaxton CS (2011) Nanoparticle therapeutics: FDA approval, clinical trials, regulatory pathways, and case study. Methods Mol Biol 726: 325-338.

[84] Davis ME (2009) The first targeted delivery of siRNA in humans via a self-assembling, cyclodextrin polymer-based nanoparticle: from concept to clinic. Mol Pharm 6: 659-668.

[85] Davis ME, Zuckerman JE, Choi CH, Seligson D, Tolcher A, et al. (2010) Evidence of RNAi in humans from systemically administered siRNA via targeted nanoparticles. Nature 464: 1067-1070.

Chapter 7

Recovery from ICH – Potential Targets

Yao Yao and Stella E. Tsirka

Additional information is available at the end of the chapter

http://dx.doi.org/10.5772/58477

1. Introduction

Intracerebral hemorrhage (ICH) is a devastating clinical event caused by rupture of blood vessels and accumulation of blood in the brain. Many disorders, including hypertensive arteriosclerosis, amyloid angiopathy, neoplasia, coagulation disorders and cerebrovascular malformations, directly or indirectly damage blood vessels in the brain and thus lead to ICH. The annual occurrence of ICH is estimated to be approximately 0.12 million in the USA and 2 million in the world. These numbers are expected to increase due to the aging of populations. Although accounting for only 15-20% of all strokes, ICH has severe clinical symptoms and poor prognosis. The 1-year survival rate of ICH is estimated to be 38% and long-term physical and mental disability is found in more than 90% of the survivors. Sadly, there is no effective treatment for ICH. Currently, primary supportive care and risk factor control are the main therapy for ICH in clinics. Thus, research and development of effective reagents to treat ICH is extremely urgent. In this chapter, we first introduce the anatomy and biology of the blood brain barrier. Then the pathophysiology and animal models of ICH are reviewed. Furthermore, we summarize the potential therapeutic targets for ICH.

2. Blood brain barrier

One unique feature about the blood vessels in the brain is the presence of the blood-brain barrier (BBB). BBB is a natural barrier that separates the central nervous system (CNS) from the circulation [1]. Under physiological conditions, the BBB prevents the entrance of blood cells and large molecules into the brain, but allows the uptake of nutrients and hormones from the blood, maintaining the homeostasis of CNS microenvironment [1, 2]. Under pathological conditions, the integrity of BBB is compromised and blood components leak into the brain, contributing to the progress of diseases [3-12]. At the cellular level, the BBB consists of brain microvascular endothelial cells (BMECs), astrocytes, pericytes, neurons, microglia, and the non-cellular component-basement membrane [13] (Figure 1).

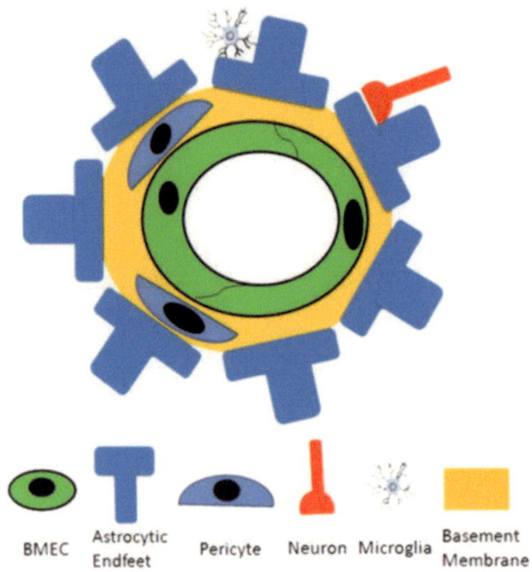

Figure 1. Schematic illustration of BBB. The BBB is composed of cellular and non-cellular components. Cellular components include BMECs, perycites, astrocytic endfeet, neurons, and microglia. Non-cellular components includes the basement membrane.

a. BMECs

Endothelial cells in the CNS, BMECs, are different in many ways from the ones in the periphery. First, BMECs have more mitochondria, lower pinocytotic activity, and little-to-no fenestrations. Second, the endothelium in the brain and the spinal cord is 50-100 times tighter than that in the rest of the body [14]. In the CNS, BMECs connect to each other via tight junctions, which are unique structures that confer impermeability to the BBB. Two types of proteins are found at tight junctions: transmembrane proteins, including occludin and claudins, and cytoplasmic accessory proteins, including zonula occluden-1, 2, 3 (ZO-1, 2, 3) and cingulin [15, 16]. The transmembrane proteins seal gaps between adjacent cells, decreasing intercellular permeability [17, 18], whereas cytoplasmic accessory proteins link transmembrane proteins to cortical actin-based cytoskeleton, enabling strict regulation of the distribution of tight junction proteins (TJP) [19, 20].

Besides intercellular transportation, intracellular transportation is another way to regulate BBB permeability [11, 21-23]. Although small lipophilic molecules, such as oxygen and carbon dioxide, can diffuse across BMECs freely [24], the transport of large hydrophilic molecules is mediated by specific transporters or receptors. Based on their subcellular distribution and functions, these transporters and receptors are divided into three groups. Group I transporters are expressed on both luminal and abluminal sides of BMECs and function to transport nutrients between the blood and brain [25, 26]. For example, glucose transporter 1 (GLUT1) transports glucose; monocarboxylate transporter 1 (MCT1) transports lactate; the L1 and y +transporters transport large neutral and cationic essential amino acids to and from the brain. Group II transporters are also expressed on both sides of BMECs, but only transport materials in one direction [27-29]. For example, transferrin and insulin receptors (TFR and IR) are expressed on both sides of BMECs. The luminal and abluminal receptors mediate endocytosis of transferrin and insulin from the blood and brain, respectively. Group III transporters are expressed on only one side of BMECs and usually mediate one-way transportation of materials [26, 30-38]. For instance, in order to remove excitatory neurotransmitter glutamate from the brain, excitatory amino acid transporters (EAATs) are exclusively expressed on the abluminal side of BMECs. Similarly, to facilitate the removal of amyloid-β from the brain, low-density lipoprotein receptor related protein 1 (LRP1) is solely expressed on the abluminal side of BMECs. Another example of such transporters is (Na^+-K^+) ATPase, which is only found on the abluminal side to regulate ion homeostasis and thus proper neuronal & synaptic functions. Additionally, multidrug resistance related protein 1 (MRP1) and P-glycoprotein (P-gp) are primarily expressed on BMEC luminal side to efflux many types of drugs from the brain. The subcellular distribution of these transporters and receptors is summarized in Figure 2.

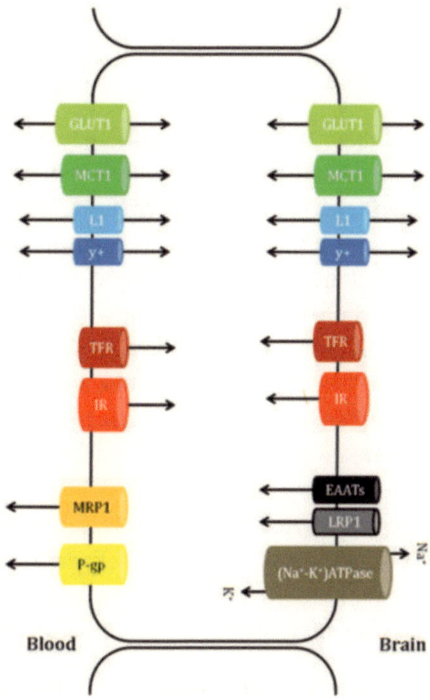

Figure 2. Major transporters and receptors expressed by BMECs. Three groups of transporters are expressed in BMECs. Group I includes GLUT1, MCT1, L1 and y+transporters, which are expressed on both luminal and abluminal sides of BMECs and transport materials bi-directionally. Group II includes TRP and IR, which are expressed on both sides of BMRCs but transport materials in one direction. Group III includes EAATs, LRP1, (Na+-K+)ATPase, MRP1 and P-gp, wich are expressed on only one side of BMECs.

b. Astrocytes

More than 30 years ago Stewart and Wiley, using xenograft experiments, demonstrated that the unique properties of BMECs, including increased mitochondria number, few pinocytotic vesicles and presence of tight junctions [39], were induced by the microenvironment of the CNS. Astrocytes, which constitute the major glial cells in the brain that cover more than 99% of the vascular surface using their extended endfeet [40, 41], have been suggested to contribute to these unique features of BMECs as well as the impermeability of BBB. Consistent with this hypothesis, temporary focal loss of astrocytes positively correlates with the compromise of BBB integrity *in vivo* [42]. Additionally, injected astrocytes have been shown to cover the blood vessels in the eye and prevent the leakage of Evans blue from the circulation system [43]. Moreover, *in vitro* culture experiments revealed that BMEC-astrocyte co-culture had a higher transendothelial electrical resistance (TEER) and less leakage of tracers, compared to BMEC monolayer [44-46]. Further mechanistic studies have demonstrated that both direct contact and astrocyte-secreted soluble factors, such as Ang1, TGF-β, GDNF and FGF2, are responsible for the impermeability of BBB [47-49]. These data suggest that astrocytes, by interacting with BMECs directly and indirectly, contribute to the unique properties of BMECs and the impermeability of BBB. Therefore, the co-culture of BMEC with astrocytes has been one of the most widely used *in vitro* BBB models, since it replicates in a petri dish the tight structures observed *in vivo*.

c. Pericytes

Discovered in 1873, pericytes are perivascular cells sandwiched between endothelial cells and astrocytic endfeet [50]. They are embedded in the basement membrane under normal conditions [1]. Pericytes cover capillaries and the degree of coverage varies depending on the species and tissue type[51]. It has been shown that the pericyte-to-endothelial ratio is 1:5 in rats, 1:4 in mice, and 1:3-4 in humans [52, 53]. In mice, this ratio is 1:1 in retina, 1:3 in brain and 1:100 in skeletal muscle vasculature [54], representing how tightly the blood vessels and their contents are confined in different tissues. Pericytes have several different developmental origins, depending on the organs they cover [51]. In the brain and thymus, pericytes arise from ectoderm-derived neural crest, whereas they differentiate from the mesothelium in the lungs, liver, and gut [51]. So far, there are no pericyte-specific markers available [51], although many cellular markers, including α-smooth muscle actin (SMA), PDGFRβ, Desmin, CD13, NG2, and RGS-5, have been used to identify pericytes, primarily in combination, as none of these markers is exclusive for these cells (pericytes share markers with myofibroblasts, vascular smooth muscle cells and neuronal progenitors [51]). It should be noted that the expression of these markers is high dependent on the differentiation stage of pericytes.

The main functions of pericytes include BBB regulation, vascular development and injury repair [52, 55, 56]. Here we focus on BBB regulation. It has been shown that pericyte-deficient mice have compromised BBB and pericyte coverage positively correlates with the tightness of tight junction [11, 57, 58]. Additionally, pericytes migrate away from capillaries, decreasing their coverage, under pathological conditions, such as hypoxia and traumatic brain injury [59, 60]. These data suggest that pericytes play a critical role in BBB integrity and maintenance.

Mechanistic studies demonstrate that BBB breakdown in pericyte-deficient mice is due to diminished expression of BBB-specific genes in endothelial cells and lack of polarity in astrocytic endfeet [58]. Consistent with these data, adding pericytes to BMEC-astrocyte co-culture system significantly enhanced TEER and decreased the leakage of tracers [61, 62]. Further studies showed that the function of pericytes on BBB integrity is also dependent on the differentiation stage of pericytes [63]. TGF-β treated pericytes, which are further differentiated SMA^{+}pericytes, compromise BBB integrity. On the contrary, b-FGF treated pericytes, which are less differentiated SMA^{-}pericytes, maintain impermeability of BBB. Altogether, these data suggest that pericytes is a key regulator of the BBB integrity. Nowadays, BMEC-pericyte-astrocyte triple-culture is becoming more and more popular in BBB research.

d. Neurons

In the human brain, the number of neurons and capillaries is estimated to be the same [64]. Both BMECs and astrocytic processes are directly innervated by noradrenergic, serotonergic, cholinergic, and GABA-ergic neurons [65-71]. The fact that local neuronal activity and metabolism regulate cerebral blood flow (neurovascular coupling) suggests that neurons may regulate BBB permeability through modulating BMEC and astrocyte function [72]. Consistent with these data, adding neurons to *in vitro* BBB models significantly increases the tightness of the BBB [73]. However, the exact mechanism underlying how neurons contribute to the BBB integrity is still elusive. Many studies focus on such mechanisms.

e. Microglia

Microglia, the brain resident immune competent cells, account for 10-20% of glial cells in the brain [74, 75]. Fate mapping studies suggest that they originate from Myb-independent, FLT3-independent, but PU.1-dependent myeloid progenitors that express colony stimulating factor 1 receptor (Csf1R) at embryonic day 8.5 [76-80]. Under physiological conditions, microglia have a ramified morphology, characterized by a small cell body and many long/thin dynamic processes [75]. By extending and retracting these dynamic processes, microglia survey the changes of microenvironment in the brain [75]. Once an insult is identified, microglia quickly undergo a process collectively termed activation, which involves changes to ameboid morphology. Activated microglia migrate to the site of injury, proliferate locally, secrete pro-and anti-inflammatory cytokines, and remove cellular debris by phagocytosis [74, 81-83]. Microglia play a dual role in the brain. On one hand, they contribute to neurite growth and neuronal survival by clearing cell debris and releasing neurotrophic factors [84-86], such as neurotrophin-3 and brain-derived neurotrophic factor. On the other hand, microglia secrete high levels of pro-inflammatory cytokines, including TNF-α and IL-1β, promoting neuronal death. The former (neuroprotective microglia) display anti-inflammatory properties and are called M2 cells, similar to the nomenclature of macrophages. The latter, secreting pro-inflammatory cytokines, exhibit neurotoxic behaviors and are called M1 microglia. Which role they play is highly dependent on the timing after injury and the type of injury. Since microglia are close to other components of the BBB in the brain, they may regulate BBB integrity either by directly interacting with the blood vessels, or indirectly through interaction with BMECs, astrocyte endfeet, or pericytes [87]. Interestingly, microglial activation has been reported to both compromise and restore BBB integrity [88, 89]. This discrepancy could be explained by different injury models and different timing after injury. More work is needed to answer the question how microglia regulate BBB integrity.

f. Basement Membrane (BM)

BM is a 3-dimensional network composed of extracellular matrix (ECM) proteins, including collagens, laminins, heparin sulfate proteoglycans, and nidogens [47, 90]. The formation of this network involves polymerization and cross-link of these ECM proteins [90, 91]. At the BBB, BMECs generate a vascular BM and astrocytes generate a parenchymal BM [92, 93]. The vascular and parenchymal BM is usually indistinguishable at capillaries [1]. However, at the post-capillary venules, the two BMs are separated by perivascular space where cerebrospinal fluid drains, and where antigen-presenting cells can be found [1]. Both BM layers have the same composition except that in the vascular BM laminin-α4 and-α5 are predominantly present [93], whereas in the parenchymal BM laminins-α1 and-α2 are the main components [92-94].

Accumulating evidence suggests that loss of BM results in disruption of BBB, probably due to the loss of a physical barrier at the BMEC-astrocyte interface and/or lack of signaling from ECM molecules [95-99]. Individual ECM proteins, including laminin, collagen type IV, and fibronectin, have been shown to increase the TEER of BMECs *in vitro* [100]. Using laminin conditional knockout mice, we have shown that astrocytic laminin maintains BBB integrity by preventing pericyte differentiation from the resting stage to the contractile stage [101]. In addition, laminin α5 and dystroglycan, a major receptor for ECM proteins, have been found to negatively correlate with the infiltration of leukocytes in the brain [93]. These data suggest that BM plays a crucial role in BBB regulation. Future studies are expected to focus on the roles of individual ECM proteins in BBB integrity. Understanding how these ECM proteins affect individual BBB components and BBB integrity would significantly enhance our knowledge on BBB and potentially pave the way for the treatment of many neurological disorders.

3. Pathophysiology

When ICH occurs, blood leaks into the brain parenchyma, leading to the formation of hematoma, which quickly increases intracranial pressure. The accumulated blood and high intracranial pressure cause immediate primary damage to the brain. This initial injury is followed by secondary damage mainly resulting from inflammatory responses [102, 103]. The exposure of brain parenchyma to blood proteins (e.g., proteases and hemoglobin) and cells (red blood cells and leukocytes) results in activation of microglia, and the secretion of pro-inflammatory cytokines/chemokines [104, 105], including TNF-α, IL-1β, and MCP1/CCL2. These inflammatory mediators, by forming a concentration gradient, activate and attract more microglia and other inflammatory cells to the injury site [106]. These cells then accumulate around the hematoma, forming a barrier to prevent the spread of injury to other sites. The released pro-inflammatory cytokines/chemokines and possibly activated microglia also act on BMECs, pericytes and astrocytes, leading to compromise of BBB integrity. Through the disrupted BBB, peripheral leukocytes infiltrate into the brain. The infiltrated leukocytes together with activated microglia produce more pro-inflammatory mediators, which induce cell death in the penumbra area [107, 108]. In addition, hemolysis of red blood cells causes iron deposition in the brain parenchyma and subsequent lipid peroxidation [109]. Free radicals generated during lipid peroxidation also lead to cell death and contribute to ICH-induced brain injury [110, 111]. With the progress of disease, microglia and infiltrated leukocytes change their gene expression profile from pro-inflammatory to anti-inflammatory and clear up the dead cells via phagocytosis [110, 112]. The clearance of cell debris finally leads to the resolution of the hematoma and repair of damaged tissue. At this stage, the activated inflammatory cells revert to a resting state again. Due to the limited regenerative ability of neurons, however, most neurological functions cannot be recovered, which explains the high extent of disability after ICH.

4. ICH animal models

To study ICH and eventually cure this disease, several ICH animal models have been developed, including collagenase ICH model, whole blood ICH model, and the spontaneous ICH model. Although these models have been widely used in ICH research, none of them fully replicates the pathology of ICH in human patients. Here we briefly discuss the advantages and disadvantages of these models.

a. Collagenase ICH Model

This model utilizes the enzymatic activity of collagenase, a bacterial enzyme. After injection into the brain, collagenase induces rupture of blood vessels by degrading collagen IV, a component of the blood vessel wall [103-105]. The rupture of blood vessels then induces the formation of hematoma and other pathological alterations. There are many advantages of this model. First, ICH induced by collagenase injection is very reliable and reproducible. The size and location of hematoma reported by different laboratories across the world are comparable [112-115]. Second, the location of hematoma can be controlled depending on the site of injection. Third, this model is very simple and fast. ICH can be induced within hours after collagenase injection. Due to these advantages, collagenase ICH model has become one of the most popular animal models for ICH research. This model, however, also has a few disadvantages. One of the most significant drawbacks is that it introduces collagenase, a bacterial enzyme, into the mammalian brain. This enzyme degrades ECM proteins in the brain, affects BBB integrity, and modifies inflammatory or immune responses, all of which may affect ICH progress [105, 116, 117]. Another disadvantage of this model is that it does not replicate the vascular challenges usually seen before the onset of ICH in patients, such as hypertension and atherosclerosis. Mice lacking these vascular injuries may have different disease progress and/or recovery patterns, which makes it difficult to interpret data generated using this ICH model.

b. Whole Blood ICH Model

The whole blood model involves injection of blood from the same animal or a donor into the brain. The injected blood induces secondary pathological changes observed in human patients. Unlike the collagenase ICH model, this model does not introduce exogenous enzymes. The application of this ICH model, however, is circumvented by its three major disadvantages. First, the whole blood ICH model lacks pathological changes in blood vessels. The vascular challenges and rupture of vasculature cannot be replicated in this model. Second, this model is less reproducible than collagenase ICH model. The size and location of hematoma vary depending on different laboratories. Third, the shape of hematoma is different from that found in human patients. Hematoma formed in whole blood ICH model is usually umbrella-shaped and narrower slit-like [118]. This unique shape is probably caused by high pressure-induced rapid distribution of blood along white matter tracts and/or corpus callosum after injection. A way to get around this problem is used in bigger animals, like pigs, where a space/balloon forming initial injection is followed by the injection of the homologous whole blood.

c. Spontaneous ICH Model

To better replicate the pathological changes observed in human patients, a spontaneous ICH model has been developed in rodents [119]. This new model induces ICH through acute hypertension, the most common etiology of hemorrhage in humans. In this model, animals are administered with NG-nitro-L-arginine methyl ester (L-NAME) and angiotensin II to induce hypertension. The injection of angiotensin II causes surges of blood pressure, which eventually lead to rupture of blood vessels and thus ICH. This spontaneous ICH model replicates most pathological alterations observed in human patients. However, the time it takes to induce ICH is relatively long (2-4 weeks), the location of the ICH varies, and the reproducibility still needs further investigation.

5. Targets for ICH treatments

ICH is a devastating clinical event. Sadly, no effective treatments are available at present. Current therapy is mainly supportive care [120, 121]. Due to the pivotal role of inflammatory responses in ICH development, anti-inflammatory strategies have been explored by many laboratories. Here we review a few anti-inflammatory targets with therapeutic potential in ICH: microglial activation, leukocyte infiltration, cytokines/chemokines, protease activation, and reactive oxygen species (ROS) production. In addition, stem cell therapy is also discussed briefly.

a. Microglial Activation

Microglia are one of the first cell types that respond to ICH. In collagenase ICH model, microglial activation starts at 1 hour [102, 122], peaks at 3-7 days [104, 105, 115, 123], and returns to a resting state again by 3-4 weeks after the onset of ICH [124, 125]. A similar time course of microglial activation is observed in whole blood ICH model [122, 124, 125]. Since activated microglia contribute to the amplification of inflammatory responses and cell death by secreting chemotactic cytokines and cytotoxic mediators, including proteases and ROS [102, 103, 112,

115], inhibition of microglial activation has been proposed as a therapeutic strategy for ICH. It has been shown that pre-or post-treatment with the tri-peptide microglia/macrophage inhibitory factor (MIF, Thr-Lys-Pro) significantly inhibited microglial activation, reduced injury size and improved neurological function [104, 105]. Consistently with this report, inhibiting microglial activation with neuroprotectant minocycline in both collagenase and whole blood ICH models protected BBB integrity, decreased brain edema, and improved functional recovery, although neuronal death remained changed [126-130]. These data support that inhibition of microglial activation is beneficial. However, there is also evidence suggesting that long-term inhibition of microglial activation is detrimental [104, 115]. Given that activated microglia also contribute to the clearance of cell debris and recovery at late stage, inhibition of microglial activation should be limited to the early stage. The question then becomes how to define early and late stages after ICH? Definition of these stages would significantly improve the outcome of ICH treatments.

b. Leukocyte Infiltration

Leukocytes infiltrate into the brain through the compromised BBB and modulate the progress and/or recovery of ICH [102, 112]. Among all the subtypes of leukocytes, neutrophils are the earliest ones to infiltrate into the brain after ICH. In both collagenase and whole blood ICH models in rodents, neutrophil infiltration starts at approximately 4 hours and peaks at 3 days after the onset of ICH [102, 115, 124, 131, 132]. These cells promote cell death and brain damage by producing ROS and pro-inflammatory mediators [107, 108], and usually die within 2 days in the brain. Mice deficient for CD18, a subunit of β2 integrin indispensable for leukocyte infiltration, demonstrated reduced brain edema and mortality as well as decreased leukocyte number in the brain after collagenase injection [133]. In human postmortem brains, leukocyte infiltration was also observed within hours after ICH [134, 135]. Furthermore, leukocyte counts in blood have been found to positively correlate with injury size in ICH patients [136]. Therefore, high leukocyte counts together with other factors have been used to predict early clinical outcome in ICH patients [137, 138]. Currently, no anti-leukocyte infiltration strategies have been investigated in ICH models. Obtaining such data may facilitate the research and development of novel reagents targeting leukocyte infiltration.

c. Cytokines/Chemokines

During ICH, activated microglia and infiltrated leukocytes produce high levels of inflammatory cytokines/chemokines, which mediate the secondary damage to the brain. In both rodents and humans, pro-inflammatory cytokines, including TNF-α and IL-1β, are transiently upregulated in the peri-hematomal region [106, 139]. In addition, chemokines and chemokine receptors that mediate leukocyte extravasation, including CCL2-4, IL-8, CXCL5, and CCR1-2, are also increased/activated [139, 140]. These data suggest that targeting cytokine/chemokine signaling may be a therapeutic strategy for ICH. In collagenase ICH model, we have found that mice deficient for CCL2 or its receptor CCR2 have a mild but delayed disease progression [115]. In CCL2$^{-/-}$or CCR2$^{-/-}$mice, hematoma was smaller at day 1 post injury (dpi 1) but larger at subsequent times (dpi 7 and 14 [115]), indicating a delayed recovery. Consistent with the crucial role of CCL2-CCR2 system in microglial activation/migration, limited numbers of microglia were observed at dpi 1 in both knockout mice [115]. At dpi 3 and 7, however, the number of microglia in the knockout mice far exceeded those in control animals [115], suggesting that CCL2-CCR2 independent alternative signaling recruited microglia in the knockout mice. The infiltration of neutrophils was also ablated in both knockout mice at dpi 1 and 3, echoed by the smaller hematoma size early after injury [115]. In addition, at dpi 7 the expression of inducible nitric oxide synthase (iNOS) decreased in controls compared to earlier time-points, but remained high in the mutant mice, indicating that lack of CCL2-CCR2 signaling produces more ROS. Moreover, brain edema, neuronal loss and neurological function followed similar trends over time as that of hematoma size [115]. Altogether, these data suggest that inhibiting CCL2-CCR2 signaling early after ICH is neuroprotective, whereas long-term inhibition delays the recovery. Future work should focus on developing the best CCL2-CCR2 inhibition regimen for ICH patients.

d. Protease Activation

ICH activates many proteases, including matrix metalloproteinases (MMPs). MMPs are a group of zinc-dependent proteases actively involved in extracellular remodeling and neuro-inflammation. Under physiological conditions, low levels of inactive MMPs are found in the brain. These MMPs, however, are dramatically up-regulated and activated when ICH occurs [112, 141]. We and others have demonstrated that collagenase quickly activates and up-regulates the expression of MMP-2,-3,-9, and-12 in rodents [112, 142]. Activation of MMP-9 has also been described in other ICH models [143-145]. In human ICH patients, blood MMP-9 level has been reported to correlate with BBB integrity, hematoma size, edema of the penumbra area, and neurological function [138, 146, 147], whereas blood MMP-3 levels have been found to associate with mortality [148]. Additionally, higher level of MMP-9 was detected in the peri-hematomal region in postmortem human brains [113, 149]. These data suggest that modulation of MMP activity may have therapeutic effect in ICH. Consistent with this hypothesis, mice lacking MMP-3,-9, or-12 are partially protected from ICH [141, 144, 150]. In addition, the therapeutic effect of MMP inhibitors has also been investigated. GM6001, a broad-spectrum MMP inhibitor, has been found to be neuroprotective in both collagenase and whole blood ICH models in mice [132, 151]. Similar results have been noted for BB-1101, another broad-spectrum MMP inhibitor [152]. However, both neuroprotective and detrimental roles have been reported for MMP inhibitor BB-94, depending on the animal models used [153-155]. Besides its inhibitory effect on microglial activation, minocycline also functions as a MMP inhibitor [126]. There is evidence suggesting that minocycline reduces TNF-α level and brain edema without affecting neuronal loss [127, 156], when administered 6 hours after ICH. Together, these data suggest that MMPs, especially MMP-9, play a detrimental role in ICH, and that MMP inhibitors may be used, alone or in combination with other medicine, to treat ICH.

e. ROS Production

One of the main pathological changes of ICH is the accumulation of blood in the brain. The hemolysis of extravasated red blood cells leads to degradation of hemoglobin and deposition of iron in the brain [109]. In rats, a 3-fold increase of non-heme iron was found after ICH [109].

Accumulated iron has been shown to induce oxidative stress by formation of free radicals, mediate secondary inflammatory injury, and contribute to brain atrophy and neurological deficits after ICH [157, 158]. In human patients with spontaneous ICH, blood ferritin level associates with brain edema in peri-hematomal region [159]. In addition, iron level in the hematoma also correlates with brain edema in peri-hematomal area [160]. These data suggest that iron deposition contributes to brain damage, and that removing the deposited iron may be an appropriate therapeutic approach. Consistent with this hypothesis, 2, 2'-dipyridyl, a lipid-soluble iron chelator, has been shown to be beneficial in both the collagenase and whole blood ICH models in mice [161]. Another iron chelator deferoxamine has shown neuroprotective effects in the whole blood ICH model in rats and piglets [162-165]. In collagenase ICH model, however, deferoxamine failed to show any beneficial effects [166], suggesting the effect of deferoxamine depends on ICH animal models. None-the-less high doses of deferoxamine are currently examined in clinical trials (starting in 2012) for the treatment of ICH.

An alternative way to treat iron-induced oxidative stress is to target antioxidant enzymes. To remove extra ROS, antioxidant enzymes, including glutathione S transferases, glutathione peroxidase, and glutamate-cysteine ligase, are up-regulated. The key transcription factor that controls the expression of these antioxidant enzymes is Nrf2 [167]. Nrf2 is expressed in neuronal and glial cells in the brain. Activation of Nrf2 has been shown to be neuroprotective both *in vitro* and *in vivo* [168, 169]. Additionally, mice deficient for Nrf2 showed more severe neurological deficits compared to wild-type mice in both collagenase and whole blood ICH models [170, 171]. Paralleled with neurological deficits, enhanced ROS production and leukocyte infiltration were observed in $Nrf2^{-/-}$mice [170, 171]. More importantly, sulforaphane, an Nrf2 inducer, has been reported to improve neurological deficits in mice when administered 30 minutes after ICH [170]. Together, these data suggest that Nrf2 is a target with therapeutic potential.

f. Stem Cell Therapy

ICH induces neuronal death and loss of neurological function. Multipotent stem cells with the ability to differentiate into neurons are a potential therapy for ICH. It has been reported that human neural stem cells are able to differentiate into neurons and astrocytes, and thus improve neurological function after intravenous injection in collagenase ICH model [172]. Stem cell therapy is relatively new and more work is needed before it can be used in ICH patients. For example, the route, dose and timing of stem cell injection need to be optimized; the differentiation, proliferation and integration of stem cells *in vivo* should be investigated; and the side effects of stem cell administration must be examined.

6. Summary

Accumulating evidence suggests that the secondary inflammatory responses play a critical role in the development of ICH, indicating that the molecular mechanism of inflammation is an ideal target for the therapy of ICH. As discussed, many pathways, including microglia activation, leukocyte infiltration, cytokine/chemokine secretion, protease activation, and ROS production, have been explored, and several compounds showed significant potential in the treatment of ICH. However, it should be noted that the animal models used in the studies are not perfect, which limits the interpretation of experimental data. Thus, other models and human samples should be used to confirm the results before they are used in patients.

Author details

Yao Yao[1,2] and Stella E. Tsirka[1*]

*Address all correspondence to: stella@pharm.stonybrook.edu

1 Program in Molecular and Cellular Pharmacology, Department of Pharmacological Sciences, Stony Brook University, Stony Brook, NY, USA

2 Laboratory of Neurobiology and Genetics, The Rockefeller University, New York, NY, USA

References

[1] Engelhardt B, Ransohoff RM. Capture, crawl, cross: the T cell code to breach the blood-brain barriers. Trends Immunol 2012; 33:579-589.

[2] Cestelli A, Catania C, D'Agostino S, Di Liegro I, Licata L, Schiera G, Pitarresi GL, Savettieri G, De Caro V, Giandalia G, Giannola LI. Functional feature of a novel model of blood brain barrier: studies on permeation of test compounds. J Control Release 2001; 76:139-147.

[3] Garbuzova-Davis S, Rodrigues MC, Hernandez-Ontiveros DG, Louis MK, Willing AE, Borlongan CV, Sanberg PR. Amyotrophic lateral sclerosis: a neurovascular disease. Brain Res 2011; 1398:113-125.

[4] Jain RK, di Tomaso E, Duda DG, Loeffler JS, Sorensen AG, Batchelor TT. Angiogenesis in brain tumours. Nat Rev Neurosci 2007; 8:610-622.

[5] Zlokovic BV. Neurovascular mechanisms of Alzheimer's neurodegeneration. Trends Neurosci 2005; 28:202-208.

[6] Selkoe DJ. Alzheimer disease: mechanistic understanding predicts novel therapies. Ann Intern Med 2004; 140:627-638.

[7] Kortekaas R, Leenders KL, van Oostrom JC, Vaalburg W, Bart J, Willemsen AT, Hendrikse NH. Blood-brain barrier dysfunction in parkinsonian midbrain in vivo. Ann Neurol 2005; 57:176-179.

[8] Benchenane K, Lopez-Atalaya JP, Fernandez-Monreal M, Touzani O, Vivien D. Equivocal roles of tissue-type plasminogen activator in stroke-induced injury. Trends Neurosci 2004; 27:155-160.

[9] Moskowitz MA, Lo EH, Iadecola C. The science of stroke: mechanisms in search of treatments. Neuron 2010; 67:181-198.

[10] Iadecola C. The overlap between neurodegenerative and vascular factors in the pathogenesis of dementia. Acta Neuropathol 2010; 120:287-296.

[11] Bell RD, Winkler EA, Sagare AP, Singh I, LaRue B, Deane R, Zlokovic BV. Pericytes control key neurovascular functions and neuronal phenotype in the adult brain and during brain aging. Neuron 2010; 68:409-427.

[12] Shlosberg D, Benifla M, Kaufer D, Friedman A. Blood-brain barrier breakdown as a therapeutic target in traumatic brain injury. Nat Rev Neurol 2010; 6:393-403.

[13] Guillemin GJ, Brew BJ. Microglia, macrophages, perivascular macrophages, and pericytes: a review of function and identification. J Leukoc Biol 2004; 75:388-397.

[14] Abbott NJ. Astrocyte-endothelial interactions and blood-brain barrier permeability. J Anat 2002; 200:629-638.

[15] Citi S, Cordenonsi M. Tight junction proteins. Biochim Biophys Acta 1998; 1448:1-11.

[16] Huber JD, Egleton RD, Davis TP. Molecular physiology and pathophysiology of tight junctions in the blood-brain barrier. Trends Neurosci 2001; 24:719-725.

[17] Forster C. Tight junctions and the modulation of barrier function in disease. Histochem Cell Biol 2008; 130:55-70.

[18] Ogunrinade O, Kameya GT, Truskey GA. Effect of fluid shear stress on the permeability of the arterial endothelium. Ann Biomed Eng 2002; 30:430-446.

[19] Brown RC, Davis TP. Hypoxia/aglycemia alters expression of occludin and actin in brain endothelial cells. Biochem Biophys Res Commun 2005; 327:1114-1123.

[20] Huber JD, Hau VS, Borg L, Campos CR, Egleton RD, Davis TP. Blood-brain barrier tight junctions are altered during a 72-h exposure to lambda-carrageenan-induced inflammatory pain. Am J Physiol Heart Circ Physiol 2002; 283:H1531-1537.

[21] MacAulay N, Zeuthen T. Water transport between CNS compartments: contributions of aquaporins and cotransporters. Neuroscience 2010; 168:941-956.

[22] Madri JA. Modeling the neurovascular niche: implications for recovery from CNS injury. J Physiol Pharmacol 2009; 60 Suppl 4:95-104.

[23] Miller DS. Regulation of P-glycoprotein and other ABC drug transporters at the blood-brain barrier. Trends Pharmacol Sci 2010; 31:246-254.

[24] Ballabh P, Braun A, Nedergaard M. The blood-brain barrier: an overview: structure, regulation, and clinical implications. Neurobiol Dis 2004; 16:1-13.

[25] Redzic Z. Molecular biology of the blood-brain and the blood-cerebrospinal fluid barriers: similarities and differences. Fluids Barriers CNS 2011; 8:3.

[26] Zlokovic BV. Neurovascular pathways to neurodegeneration in Alzheimer's disease and other disorders. Nat Rev Neurosci 2011; 12:723-738.

[27] Banks WA. Blood-brain barrier as a regulatory interface. Forum Nutr 2010; 63:102-110.

[28] Nishijima T, Piriz J, Duflot S, Fernandez AM, Gaitan G, Gomez-Pinedo U, Verdugo JM, Leroy F, Soya H, Nunez A, Torres-Aleman I. Neuronal activity drives localized blood-brain-barrier transport of serum insulin-like growth factor-I into the CNS. Neuron 2010; 67:834-846.

[29] Pardridge WM. Blood-brain barrier delivery. Drug Discov Today 2007; 12:54-61.

[30] O'Kane RL, Martinez-Lopez I, DeJoseph MR, Vina JR, Hawkins RA. Na(+)-dependent glutamate transporters (EAAT1, EAAT2, and EAAT3) of the blood-brain barrier. A mechanism for glutamate removal. J Biol Chem 1999; 274:31891-31895.

[31] Bell RD, Sagare AP, Friedman AE, Bedi GS, Holtzman DM, Deane R, Zlokovic BV. Transport pathways for clearance of human Alzheimer's amyloid beta-peptide and apolipoproteins E and J in the mouse central nervous system. J Cereb Blood Flow Metab 2007; 27:909-918.

[32] Deane R, Wu Z, Sagare A, Davis J, Du Yan S, Hamm K, Xu F, Parisi M, LaRue B, Hu HW, et al. LRP/amyloid beta-peptide interaction mediates differential brain efflux of Abeta isoforms. Neuron 2004; 43:333-344.

[33] Shibata M, Yamada S, Kumar SR, Calero M, Bading J, Frangione B, Holtzman DM, Miller CA, Strickland DK, Ghiso J, Zlokovic BV. Clearance of Alzheimer's amyloid-ss(1-40) peptide from brain by LDL receptor-related protein-1 at the blood-brain barrier. J Clin Invest 2000; 106:1489-1499.

[34] Neuwelt EA, Bauer B, Fahlke C, Fricker G, Iadecola C, Janigro D, Leybaert L, Molnar Z, O'Donnell ME, Povlishock JT, et al. Engaging neuroscience to advance translational research in brain barrier biology. Nat Rev Neurosci 2011; 12:169-182.

[35] ElAli A, Hermann DM. ATP-binding cassette transporters and their roles in protecting the brain. Neuroscientist 2011; 17:423-436.

[36] Abuznait AH, Kaddoumi A. Role of ABC transporters in the pathogenesis of Alzheimer's disease. ACS Chem Neurosci 2012; 3:820-831.

[37] Soontornmalai A, Vlaming ML, Fritschy JM. Differential, strain-specific cellular and subcellular distribution of multidrug transporters in murine choroid plexus and blood-brain barrier. Neuroscience 2006; 138:159-169.

[38] Miller DS, Bauer B, Hartz AM. Modulation of P-glycoprotein at the blood-brain barrier: opportunities to improve central nervous system pharmacotherapy. Pharmacol Rev 2008; 60:196-209.

[39] Stewart PA, Wiley MJ. Developing nervous tissue induces formation of blood-brain barrier characteristics in invading endothelial cells: a study using quail--chick transplantation chimeras. Dev Biol 1981; 84:183-192.

[40] Simard M, Arcuino G, Takano T, Liu QS, Nedergaard M. Signaling at the gliovascular interface. J Neurosci 2003; 23:9254-9262.

[41] Kacem K, Lacombe P, Seylaz J, Bonvento G. Structural organization of the perivascular astrocyte endfeet and their relationship with the endothelial glucose transporter: a confocal microscopy study. Glia 1998; 23:1-10.

[42] Willis CL, Nolan CC, Reith SN, Lister T, Prior MJ, Guerin CJ, Mavroudis G, Ray DE. Focal astrocyte loss is followed by microvascular damage, with subsequent repair of the blood-brain barrier in the apparent absence of direct astrocytic contact. Glia 2004; 45:325-337.

[43] Janzer RC, Raff MC. Astrocytes induce blood-brain barrier properties in endothelial cells. Nature 1987; 325:253-257.

[44] Cohen-Kashi Malina K, Cooper I, Teichberg VI. Closing the gap between the in-vivo and in-vitro blood-brain barrier tightness. Brain Res 2009; 1284:12-21.

[45] Siddharthan V, Kim YV, Liu S, Kim KS. Human astrocytes/astrocyte-conditioned medium and shear stress enhance the barrier properties of human brain microvascular endothelial cells. Brain Res 2007; 1147:39-50.

[46] Yao Y, Tsirka SE. Truncation of monocyte chemoattractant protein 1 by plasmin promotes blood-brain barrier disruption. J Cell Sci 2011; 124:1486-1495.

[47] Cardoso FL, Brites D, Brito MA. Looking at the blood-brain barrier: molecular anatomy and possible investigation approaches. Brain Res Rev 2010; 64:328-363.

[48] Persidsky Y, Ramirez SH, Haorah J, Kanmogne GD. Blood-brain barrier: structural components and function under physiologic and pathologic conditions. J Neuroimmune Pharmacol 2006; 1:223-236.

[49] Quaegebeur A, Lange C, Carmeliet P. The neurovascular link in health and disease: molecular mechanisms and therapeutic implications. Neuron 2011; 71:406-424.

[50] Sa-Pereira I, Brites D, Brito MA. Neurovascular Unit: a Focus on Pericytes. Mol Neurobiol 2012.

[51] Armulik A, Genove G, Betsholtz C. Pericytes: developmental, physiological, and pathological perspectives, problems, and promises. Dev Cell 2011; 21:193-215.

[52] Dore-Duffy P. Pericytes: pluripotent cells of the blood brain barrier. Curr Pharm Des 2008; 14:1581-1593.

[53] Dore-Duffy P, Cleary K. Morphology and properties of pericytes. Methods Mol Biol 2011; 686:49-68.

[54] Dalkara T, Gursoy-Ozdemir Y, Yemisci M. Brain microvascular pericytes in health and disease. Acta Neuropathol 2011; 122:1-9.

[55] Dore-Duffy P, Katychev A, Wang X, Van Buren E. CNS microvascular pericytes exhibit multipotential stem cell activity. J Cereb Blood Flow Metab 2006; 26:613-624.

[56] Fisher M. Pericyte signaling in the neurovascular unit. Stroke 2009; 40:S13-15.

[57] Daneman R, Zhou L, Kebede AA, Barres BA. Pericytes are required for blood-brain barrier integrity during embryogenesis. Nature 2010.

[58] Armulik A, Genove G, Mae M, Nisancioglu MH, Wallgard E, Niaudet C, He L, Norlin J, Lindblom P, Strittmatter K, et al. Pericytes regulate the blood-brain barrier. Nature 2010.

[59] Dore-Duffy P, Owen C, Balabanov R, Murphy S, Beaumont T, Rafols JA. Pericyte migration from the vascular wall in response to traumatic brain injury. Microvasc Res 2000; 60:55-69.

[60] Gonul E, Duz B, Kahraman S, Kayali H, Kubar A, Timurkaynak E. Early pericyte response to brain hypoxia in cats: an ultrastructural study. Microvasc Res 2002; 64:116-119.

[61] Dente CJ, Steffes CP, Speyer C, Tyburski JG. Pericytes augment the capillary barrier in in vitro cocultures. J Surg Res 2001; 97:85-91.

[62] Nakagawa S, Deli MA, Kawaguchi H, Shimizudani T, Shimono T, Kittel A, Tanaka K, Niwa M. A new blood-brain barrier model using primary rat brain endothelial cells, pericytes and astrocytes. Neurochem Int 2009; 54:253-263.

[63] Thanabalasundaram G, Schneidewind J, Pieper C, Galla HJ. The impact of pericytes on the blood-brain barrier integrity depends critically on the pericyte differentiation stage. Int J Biochem Cell Biol 2011; 43:1284-1293.

[64] Zlokovic BV. The blood-brain barrier in health and chronic neurodegenerative disorders. Neuron 2008; 57:178-201.

[65] Ben-Menachem E, Johansson BB, Svensson TH. Increased vulnerability of the blood-brain barrier to acute hypertension following depletion of brain noradrenaline. J Neural Transm 1982; 53:159-167.

[66] Cohen Z, Bonvento G, Lacombe P, Hamel E. Serotonin in the regulation of brain microcirculation. Prog Neurobiol 1996; 50:335-362.

[67] Cohen Z, Molinatti G, Hamel E. Astroglial and vascular interactions of noradrenaline terminals in the rat cerebral cortex. J Cereb Blood Flow Metab 1997; 17:894-904.

[68] Hawkins BT, Davis TP. The blood-brain barrier/neurovascular unit in health and disease. Pharmacol Rev 2005; 57:173-185.

[69] Tong XK, Hamel E. Regional cholinergic denervation of cortical microvessels and nitric oxide synthase-containing neurons in Alzheimer's disease. Neuroscience 1999; 92:163-175.

[70] Vaucher E, Hamel E. Cholinergic basal forebrain neurons project to cortical microvessels in the rat: electron microscopic study with anterogradely transported Phaseolus vulgaris leucoagglutinin and choline acetyltransferase immunocytochemistry. J Neurosci 1995; 15:7427-7441.

[71] Vaucher E, Tong XK, Cholet N, Lantin S, Hamel E. GABA neurons provide a rich input to microvessels but not nitric oxide neurons in the rat cerebral cortex: a means for direct regulation of local cerebral blood flow. J Comp Neurol 2000; 421:161-171.

[72] Iadecola C. Neurovascular regulation in the normal brain and in Alzheimer's disease. Nat Rev Neurosci 2004; 5:347-360.

[73] Minami M. [Neuro-glio-vascular interaction in ischemic brains]. Yakugaku Zasshi 2011; 131:539-544.

[74] Nakajima K, Kohsaka S. Microglia: neuroprotective and neurotrophic cells in the central nervous system. Curr Drug Targets Cardiovasc Haematol Disord 2004; 4:65-84.

[75] Nimmerjahn A, Kirchhoff F, Helmchen F. Resting microglial cells are highly dynamic surveillants of brain parenchyma in vivo. Science 2005; 308:1314-1318.

[76] Buza-Vidas N, Woll P, Hultquist A, Duarte S, Lutteropp M, Bouriez-Jones T, Ferry H, Luc S, Jacobsen SE. FLT3 expression initiates in fully multipotent mouse hematopoietic progenitor cells. Blood 2011; 118:1544-1548.

[77] Schulz C, Gomez Perdiguero E, Chorro L, Szabo-Rogers H, Cagnard N, Kierdorf K, Prinz M, Wu B, Jacobsen SE, Pollard JW, et al. A lineage of myeloid cells independent of Myb and hematopoietic stem cells. Science 2012; 336:86-90.

[78] Erblich B, Zhu L, Etgen AM, Dobrenis K, Pollard JW. Absence of colony stimulation factor-1 receptor results in loss of microglia, disrupted brain development and olfactory deficits. PLoS One 2011; 6:e26317.

[79] Gomez Perdiguero E, Schulz C, Geissmann F. Development and homeostasis of "resident" myeloid cells: The case of the microglia. Glia 2013; 61:112-120.

[80] Ginhoux F, Greter M, Leboeuf M, Nandi S, See P, Gokhan S, Mehler MF, Conway SJ, Ng LG, Stanley ER, et al. Fate mapping analysis reveals that adult microglia derive from primitive macrophages. Science 2010; 330:841-845.

[81] Aloisi F. Immune function of microglia. Glia 2001; 36:165-179.

[82] Hanisch UK, Kettenmann H. Microglia: active sensor and versatile effector cells in the normal and pathologic brain. Nat Neurosci 2007; 10:1387-1394.

[83] Kim SU, de Vellis J. Microglia in health and disease. J Neurosci Res 2005; 81:302-313.

[84] Nakajima K, Tohyama Y, Kohsaka S, Kurihara T. Ceramide activates microglia to enhance the production/secretion of brain-derived neurotrophic factor (BDNF) without induction of deleterious factors in vitro. J Neurochem 2002; 80:697-705.

[85] Rabchevsky AG, Streit WJ. Grafting of cultured microglial cells into the lesioned spinal cord of adult rats enhances neurite outgrowth. J Neurosci Res 1997; 47:34-48.

[86] Elkabes S, DiCicco-Bloom EM, Black IB. Brain microglia/macrophages express neurotrophins that selectively regulate microglial proliferation and function. J Neurosci 1996; 16:2508-2521.

[87] Choi YK, Kim KW. Blood-neural barrier: its diversity and coordinated cell-to-cell communication. BMB Rep 2008; 41:345-352.

[88] Willis CL. Glia-induced reversible disruption of blood-brain barrier integrity and neuropathological response of the neurovascular unit. Toxicol Pathol 2011; 39:172-185.

[89] Nishioku T, Matsumoto J, Dohgu S, Sumi N, Miyao K, Takata F, Shuto H, Yamauchi A, Kataoka Y. Tumor necrosis factor-alpha mediates the blood-brain barrier dysfunction induced by activated microglia in mouse brain microvascular endothelial cells. J Pharmacol Sci 2010; 112:251-254.

[90] Yurchenco PD, Amenta PS, Patton BL. Basement membrane assembly, stability and activities observed through a developmental lens. Matrix Biol 2004; 22:521-538.

[91] Yurchenco PD, Patton BL. Developmental and pathogenic mechanisms of basement membrane assembly. Curr Pharm Des 2009; 15:1277-1294.

[92] Owens T, Bechmann I, Engelhardt B. Perivascular spaces and the two steps to neuroinflammation. J Neuropathol Exp Neurol 2008; 67:1113-1121.

[93] Sixt M, Engelhardt B, Pausch F, Hallmann R, Wendler O, Sorokin LM. Endothelial cell laminin isoforms, laminins 8 and 10, play decisive roles in T cell recruitment across the blood-brain barrier in experimental autoimmune encephalomyelitis. J Cell Biol 2001; 153:933-946.

[94] van Horssen J, Bo L, Vos CM, Virtanen I, de Vries HE. Basement membrane proteins in multiple sclerosis-associated inflammatory cuffs: potential role in influx and transport of leukocytes. J Neuropathol Exp Neurol 2005; 64:722-729.

[95] del Zoppo GJ. Relationship of neurovascular elements to neuron injury during ischemia. Cerebrovasc Dis 2009; 27 Suppl 1:65-76.

[96] Del Zoppo GJ, Milner R, Mabuchi T, Hung S, Wang X, Koziol JA. Vascular matrix adhesion and the blood-brain barrier. Biochem Soc Trans 2006; 34:1261-1266.

[97] Kwon I, Kim EH, del Zoppo GJ, Heo JH. Ultrastructural and temporal changes of the microvascular basement membrane and astrocyte interface following focal cerebral ischemia. J Neurosci Res 2009; 87:668-676.

[98] Lo EH, Rosenberg GA. The neurovascular unit in health and disease: introduction. Stroke 2009; 40:S2-3.

[99] Wang CX, Shuaib A. Critical role of microvasculature basal lamina in ischemic brain injury. Prog Neurobiol 2007; 83:140-148.

[100] Tilling T, Korte D, Hoheisel D, Galla HJ. Basement membrane proteins influence brain capillary endothelial barrier function in vitro. J Neurochem 1998; 71:1151-1157.

[101] Yao Y, Chen ZL, Norris EH, Strickland S. Astrocytic laminin regulates pericyte differentiation and maintains blood brain barrier integrity. Nat Commun 2014; 5:3413.

[102] Wang J, Dore S. Heme oxygenase-1 exacerbates early brain injury after intracerebral haemorrhage. Brain 2007; 130:1643-1652.

[103] Wang J, Tsirka SE. Contribution of extracellular proteolysis and microglia to intracerebral hemorrhage. Neurocrit Care 2005; 3:77-85.

[104] Wang J, Tsirka SE. Tuftsin fragment 1-3 is beneficial when delivered after the induction of intracerebral hemorrhage. Stroke 2005; 36:613-618.

[105] Wang J, Rogove AD, Tsirka AE, Tsirka SE. Protective role of tuftsin fragment 1-3 in an animal model of intracerebral hemorrhage. Ann Neurol 2003; 54:655-664.

[106] Keep RF, Xiang J, Ennis SR, Andjelkovic A, Hua Y, Xi G, Hoff JT. Blood-brain barrier function in intracerebral hemorrhage. Acta Neurochir Suppl 2008; 105:73-77.

[107] Joice SL, Mydeen F, Couraud PO, Weksler BB, Romero IA, Fraser PA, Easton AS. Modulation of blood-brain barrier permeability by neutrophils: in vitro and in vivo studies. Brain Res 2009; 1298:13-23.

[108] Nguyen HX, O'Barr TJ, Anderson AJ. Polymorphonuclear leukocytes promote neurotoxicity through release of matrix metalloproteinases, reactive oxygen species, and TNF-alpha. J Neurochem 2007; 102:900-912.

[109] Wu J, Hua Y, Keep RF, Nakamura T, Hoff JT, Xi G. Iron and iron-handling proteins in the brain after intracerebral hemorrhage. Stroke 2003; 34:2964-2969.

[110] Aronowski J, Hall CE. New horizons for primary intracerebral hemorrhage treatment: experience from preclinical studies. Neurol Res 2005; 27:268-279.

[111] Xi G, Keep RF, Hoff JT. Mechanisms of brain injury after intracerebral haemorrhage. Lancet Neurol 2006; 5:53-63.

[112] Wang J, Dore S. Inflammation after intracerebral hemorrhage. J Cereb Blood Flow Metab 2007; 27:894-908.

[113] Tejima E, Zhao BQ, Tsuji K, Rosell A, van Leyen K, Gonzalez RG, Montaner J, Wang X, Lo EH. Astrocytic induction of matrix metalloproteinase-9 and edema in brain hemorrhage. J Cereb Blood Flow Metab 2007; 27:460-468.

[114] Wagner KR. Modeling intracerebral hemorrhage: glutamate, nuclear factor-kappa B signaling and cytokines. Stroke 2007; 38:753-758.

[115] Yao Y, Tsirka SE. The CCL2-CCR2 system affects the progression and clearance of intracerebral hemorrhage. Glia 2012; 60:908-918.

[116] Matsushita K, Meng W, Wang X, Asahi M, Asahi K, Moskowitz MA, Lo EH. Evidence for apoptosis after intercerebral hemorrhage in rat striatum. J Cereb Blood Flow Metab 2000; 20:396-404.

[117] Chu K, Jeong SW, Jung KH, Han SY, Lee ST, Kim M, Roh JK. Celecoxib induces functional recovery after intracerebral hemorrhage with reduction of brain edema and perihematomal cell death. J Cereb Blood Flow Metab 2004; 24:926-933.

[118] MacLellan CL, Silasi G, Auriat AM, Colbourne F. Rodent models of intracerebral hemorrhage. Stroke 2010; 41:S95-98.

[119] Wakisaka Y, Chu Y, Miller JD, Rosenberg GA, Heistad DD. Spontaneous intracerebral hemorrhage during acute and chronic hypertension in mice. J Cereb Blood Flow Metab 2010; 30:56-69.

[120] Broderick JP, Adams HP, Jr., Barsan W, Feinberg W, Feldmann E, Grotta J, Kase C, Krieger D, Mayberg M, Tilley B, et al. Guidelines for the management of spontaneous intracerebral hemorrhage: A statement for healthcare professionals from a special writing group of the Stroke Council, American Heart Association. Stroke 1999; 30:905-915.

[121] Morgenstern LB, Frankowski RF, Shedden P, Pasteur W, Grotta JC. Surgical treatment for intracerebral hemorrhage (STICH): a single-center, randomized clinical trial. Neurology 1998; 51:1359-1363.

[122] Zhao X, Sun G, Zhang J, Strong R, Song W, Gonzales N, Grotta JC, Aronowski J. Hematoma resolution as a target for intracerebral hemorrhage treatment: role for peroxisome proliferator-activated receptor gamma in microglia/macrophages. Ann Neurol 2007; 61:352-362.

[123] Wang J, Dore S. Heme oxygenase 2 deficiency increases brain swelling and inflammation after intracerebral hemorrhage. Neuroscience 2008; 155:1133-1141.

[124] Xue M, Del Bigio MR. Intracerebral injection of autologous whole blood in rats: time course of inflammation and cell death. Neurosci Lett 2000; 283:230-232.

[125] Gong C, Hoff JT, Keep RF. Acute inflammatory reaction following experimental intracerebral hemorrhage in rat. Brain Res 2000; 871:57-65.

[126] Yong VW, Wells J, Giuliani F, Casha S, Power C, Metz LM. The promise of minocycline in neurology. Lancet Neurol 2004; 3:744-751.

[127] Wasserman JK, Schlichter LC. Minocycline protects the blood-brain barrier and reduces edema following intracerebral hemorrhage in the rat. Exp Neurol 2007; 207:227-237.

[128] Xue M, Mikliaeva EI, Casha S, Zygun D, Demchuk A, Yong VW. Improving outcomes of neuroprotection by minocycline: guides from cell culture and intracerebral hemorrhage in mice. Am J Pathol 2010; 176:1193-1202.

[129] Wu J, Yang S, Xi G, Fu G, Keep RF, Hua Y. Minocycline reduces intracerebral hemorrhage-induced brain injury. Neurol Res 2009; 31:183-188.

[130] Power C, Henry S, Del Bigio MR, Larsen PH, Corbett D, Imai Y, Yong VW, Peeling J. Intracerebral hemorrhage induces macrophage activation and matrix metalloproteinases. Ann Neurol 2003; 53:731-742.

[131] Peeling J, Yan HJ, Corbett D, Xue M, Del Bigio MR. Effect of FK-506 on inflammation and behavioral outcome following intracerebral hemorrhage in rat. Exp Neurol 2001; 167:341-347.

[132] Wang J, Tsirka SE. Neuroprotection by inhibition of matrix metalloproteinases in a mouse model of intracerebral haemorrhage. Brain 2005; 128:1622-1633.

[133] Titova E, Ostrowski RP, Kevil CG, Tong W, Rojas H, Sowers LC, Zhang JH, Tang J. Reduced brain injury in CD18-deficient mice after experimental intracerebral hemorrhage. J Neurosci Res 2008; 86:3240-3245.

[134] Mackenzie JM, Clayton JA. Early cellular events in the penumbra of human spontaneous intracerebral hemorrhage. J Stroke Cerebrovasc Dis 1999; 8:1-8.

[135] Guo FQ, Li XJ, Chen LY, Yang H, Dai HY, Wei YS, Huang YL, Yang YS, Sun HB, Xu YC, Yang ZL. [Study of relationship between inflammatory response and apoptosis in perihematoma region in patients with intracerebral hemorrhage]. Zhongguo Wei Zhong Bing Ji Jiu Yi Xue 2006; 18:290-293.

[136] Bestue-Cardiel M, Martin-Martinez J, Iturriaga-Heras C, Ara-Callizo JR, Oliveros-Juste A. [Leukocytes and primary intracerebral hemorrhage]. Rev Neurol 1999; 29:968-971.

[137] Leira R, Davalos A, Silva Y, Gil-Peralta A, Tejada J, Garcia M, Castillo J. Early neurologic deterioration in intracerebral hemorrhage: predictors and associated factors. Neurology 2004; 63:461-467.

[138] Silva Y, Leira R, Tejada J, Lainez JM, Castillo J, Davalos A. Molecular signatures of vascular injury are associated with early growth of intracerebral hemorrhage. Stroke 2005; 36:86-91.

[139] Rosell A, Vilalta A, Garcia-Berrocoso T, Fernandez-Cadenas I, Domingues-Montanari S, Cuadrado E, Delgado P, Ribo M, Martinez-Saez E, Ortega-Aznar A, Montaner J. Brain perihematoma genomic profile following spontaneous human intracerebral hemorrhage. PLoS One 2011; 6:e16750.

[140] Carmichael ST, Vespa PM, Saver JL, Coppola G, Geschwind DH, Starkman S, Miller CM, Kidwell CS, Liebeskind DS, Martin NA. Genomic profiles of damage and protection in human intracerebral hemorrhage. J Cereb Blood Flow Metab 2008; 28:1860-1875.

[141] Xue M, Fan Y, Liu S, Zygun DA, Demchuk A, Yong VW. Contributions of multiple proteases to neurotoxicity in a mouse model of intracerebral haemorrhage. Brain 2009; 132:26-36.

[142] Xue M, Yong VW. Matrix metalloproteinases in intracerebral hemorrhage. Neurol Res 2008; 30:775-782.

[143] Lu A, Tang Y, Ran R, Ardizzone TL, Wagner KR, Sharp FR. Brain genomics of intracerebral hemorrhage. J Cereb Blood Flow Metab 2006; 26:230-252.

[144] Xue M, Hollenberg MD, Yong VW. Combination of thrombin and matrix metalloproteinase-9 exacerbates neurotoxicity in cell culture and intracerebral hemorrhage in mice. J Neurosci 2006; 26:10281-10291.

[145] Wu H, Zhang Z, Li Y, Zhao R, Li H, Song Y, Qi J, Wang J. Time course of upregulation of inflammatory mediators in the hemorrhagic brain in rats: correlation with brain edema. Neurochem Int 2010; 57:248-253.

[146] Abilleira S, Montaner J, Molina CA, Monasterio J, Castillo J, Alvarez-Sabin J. Matrix metalloproteinase-9 concentration after spontaneous intracerebral hemorrhage. J Neurosurg 2003; 99:65-70.

[147] Barr TL, Latour LL, Lee KY, Schaewe TJ, Luby M, Chang GS, El-Zammar Z, Alam S, Hallenbeck JM, Kidwell CS, Warach S. Blood-brain barrier disruption in humans is independently associated with increased matrix metalloproteinase-9. Stroke 2010; 41:e123-128.

[148] Alvarez-Sabin J, Delgado P, Abilleira S, Molina CA, Arenillas J, Ribo M, Santamarina E, Quintana M, Monasterio J, Montaner J. Temporal profile of matrix metalloproteinases and their inhibitors after spontaneous intracerebral hemorrhage: relationship to clinical and radiological outcome. Stroke 2004; 35:1316-1322.

[149] Rosell A, Ortega-Aznar A, Alvarez-Sabin J, Fernandez-Cadenas I, Ribo M, Molina CA, Lo EH, Montaner J. Increased brain expression of matrix metalloproteinase-9 after ischemic and hemorrhagic human stroke. Stroke 2006; 37:1399-1406.

[150] Wells JE, Biernaskie J, Szymanska A, Larsen PH, Yong VW, Corbett D. Matrix metalloproteinase (MMP)-12 expression has a negative impact on sensorimotor function following intracerebral haemorrhage in mice. Eur J Neurosci 2005; 21:187-196.

[151] Xue M, Hollenberg MD, Demchuk A, Yong VW. Relative importance of proteinase-activated receptor-1 versus matrix metalloproteinases in intracerebral hemorrhage-mediated neurotoxicity in mice. Stroke 2009; 40:2199-2204.

[152] Rosenberg GA, Navratil M. Metalloproteinase inhibition blocks edema in intracerebral hemorrhage in the rat. Neurology 1997; 48:921-926.

[153] Grossetete M, Rosenberg GA. Matrix metalloproteinase inhibition facilitates cell death in intracerebral hemorrhage in mouse. J Cereb Blood Flow Metab 2008; 28:752-763.

[154] Lapchak PA, Chapman DF, Zivin JA. Metalloproteinase inhibition reduces thrombolytic (tissue plasminogen activator)-induced hemorrhage after thromboembolic stroke. Stroke 2000; 31:3034-3040.

[155] Pfefferkorn T, Rosenberg GA. Closure of the blood-brain barrier by matrix metalloproteinase inhibition reduces rtPA-mediated mortality in cerebral ischemia with delayed reperfusion. Stroke 2003; 34:2025-2030.

[156] Wasserman JK, Schlichter LC. Neuron death and inflammation in a rat model of intracerebral hemorrhage: effects of delayed minocycline treatment. Brain Res 2007; 1136:208-218.

[157] Gutteridge JM. Hydroxyl radicals, iron, oxidative stress, and neurodegeneration. Ann N Y Acad Sci 1994; 738:201-213.

[158] Zecca L, Youdim MB, Riederer P, Connor JR, Crichton RR. Iron, brain ageing and neurodegenerative disorders. Nat Rev Neurosci 2004; 5:863-873.

[159] Mehdiratta M, Kumar S, Hackney D, Schlaug G, Selim M. Association between serum ferritin level and perihematoma edema volume in patients with spontaneous intracerebral hemorrhage. Stroke 2008; 39:1165-1170.

[160] Lou M, Lieb K, Selim M. The relationship between hematoma iron content and perihematoma edema: an MRI study. Cerebrovasc Dis 2009; 27:266-271.

[161] Wu H, Wu T, Li M, Wang J. Efficacy of the lipid-soluble iron chelator 2, 2'-dipyridyl against hemorrhagic brain injury. Neurobiol Dis 2012; 45:388-394.

[162] Hua Y, Nakamura T, Keep RF, Wu J, Schallert T, Hoff JT, Xi G. Long-term effects of experimental intracerebral hemorrhage: the role of iron. J Neurosurg 2006; 104:305-312.

[163] Okauchi M, Hua Y, Keep RF, Morgenstern LB, Xi G. Effects of deferoxamine on intracerebral hemorrhage-induced brain injury in aged rats. Stroke 2009; 40:1858-1863.